ENZYME KINETICS

This textbook for advanced courses in enzyme chemistry and enzyme kinetics covers the field of steady state enzyme kinetics from the basic principles inherent in the Henri equation to the expressions which describe the control of multi-enzyme pathways.

Steady state kinetic equations are derived with the use of the connection matrix method, and an algorithm which can be implemented easily for computer-based derivation of the equations is developed. Throughout the book emplasis is placed on a proper interpretation of the kinetic behavior of multi-reactant enzymic reactions. The principles of analytical geometry and basic calculus are employed to analyze non-hyperbolic substrate saturation curves in terms of the mechanisms, including subunit interactions, which cause the deviation from hyperbolic curves. Sensitivity theory is employed to obtain expressions for flux and concentration control coefficients for multi-enzyme pathways. The algorithm developed can be incorporated into a computer-based program for derivation. Emphasis is placed on the interpretation of control coefficients in terms of metabolic control. Problems are included at the end of each chapter and their solutions are found at the end of the book.

ENZYME KINETICS

From diastase to multi-enzyme systems

ARTHUR R. SCHULZ
Indiana University School of Medicine

CAMBRIDGE
UNIVERSITY PRESS

Published by the Press Syndicate of the University of Cambridge
The Pitt Building, Trumpington Street, Cambridge CB2 1RP
40 West 20th Street, New York, NY 10011-4211, USA
10 Stamford Road, Oakleigh, Melbourne 3166, Australia

© Cambridge University Press 1994

First published 1994

Printed in the United States of America

Library of Congress Cataloging-in-Publication Data

Schulz, Arthur R.
    Enzyme kinetics from diastase to multi-enzyme systems / Arthur R. Schulz
      p.   cm.
    ISBN 0-521-44500-0. – ISBN 0-521-44950-2 (pbk.)
    1. Enzyme kinetics.   I. Title.
    QP601.3.S38   1994   94-28309
    574.19′25 – dc20   CIP

A catalog record for this book is available from the British Library

ISBN 0-521-44500-0 Hardback
ISBN 0-521-44950-2 Paperback

Contents

Preface

One goal of this textbook is to provide the reader with an orderly development of steady state enzyme kinetics from the early formulations, through analysis of the reaction sequence of multi-reactant enzymes, to the analysis of non-hyperbolic enzyme kinetics, and finally to the control of multi-enzyme systems. The material included in this book has formed the basis of lectures on enzyme kinetics which have been given as a portion of a course on enzyme chemistry. It is hoped that it will be useful not only to the reader who is enrolled in a formal course in enzyme kinetics or enzyme chemistry, but also to readers who wish to familiarize themselves with enzyme kinetics in a self-study program, and also to the readers who wish to review the principles of steady state enzyme kinetics.

The book contains numerous equations, but neither the equations nor the derivations of the equations constitute the primary objective. Rather, it is crucial that the information contained in an equation be correlated correctly with the kinetic behavior of the enzyme. Hence, it is the kinetic behavior of the enzyme which mandates the structure of the rate equation. The task which is presented to the enzyme kineticist is to visualize the enzyme model which is consistent with the rate equation.

The reader will note that there are few references to individual enzymes in this textbook. A deliberate objective has been to present the fundamentals of enzyme kinetics in **general** terms rather than in terms of specific enzymes. The basis for this approach is the conviction that an objective investigation of the kinetic behavior of an enzyme-catalyzed reaction should be pursued in a manner which is cognizant of basic principles rather than an attempt to 'fit' the data obtained with one enzyme to the behavior of some other enzyme. The need to impose a realistic limit on the size of this textbook has led to the omission of some important materials, for example, the derivation of rate equations for enzyme-catalyzed reactions based on stochastic principles [J. Ninio, *Proc. Natl. Acad. Sci.* USA **84**: 663 (1987); A. K. Mazur, *J. Theor. Biol.* **148**: 229 (1991)]. Likewise, the structural approach to metabolic control theory developed by Reader and Mazat [C. Reder, *J. Theor. Biol.* **135**: 175 (1988)] is not included. The omission of these and other important topics from this textbook reflects only the limitation of space. It is hoped that this textbook will provide sufficient background to motivate the reader to study the foregoing papers as well as other valuable publications.

I am indebted to Dr. William F. Bosron, Dr. Robert Eisenthal, Dr. David M. Giobson and Dr. Robert A. Harris for their willingness to read the manuscript of this book. I deeply appreciate their comments. I also wish to thank Dr. Robin C. Smith of Cambridge University Press for his helpful suggestions during the preparation of the manuscript. Finally, I acknowledge those who have contributed so much to the inspiration and completion of this book, namely, my parents, who nurtured and guided me in my early life, my wife, Marian, who has loved and encouraged me, the teachers who taught and challenged me and the students who questioned and stimulated me.

Part One

Basic steady state enzyme kinetics

1

Derivation of a rate equation

Enzymes do not make reactions take place, they stimulate the rate at which reactions do take place. Any chemical reaction which proceeds in the presence of an enzyme will also proceed in the absence of the enzyme but at a *much* slower rate. Enzymes catalyze the rate of chemical reactions by lowering the activation energy of the reaction, and they do this in a manner which is highly specific for the reactants of the reaction. It was realized very early in the study of enzyme action that meaningful studies of enzyme action would, of necessity, involve the study of the kinetic behavior of the chemical reaction in the presence of the appropriate enzyme. It is still true that if one understands the kinetic behavior of the enzyme-catalyzed reaction, one also understands much about the mechanism of the enzymic reaction. This requires the investigation of the kinetic behavior of the enzymic reaction under conditions which are defined meticulously. Within the framework of this text, this will imply under steady state conditions. *Steady state*, as it applies to enzyme kinetics will be defined in this chapter and in chapter 2.

1.1 The role of 'diastase' in the early development of a theory

The enzyme-catalyzed hydrolysis of sucrose played an important role in the early development of a suitable equation to explain the kinetic behavior of enzyme-catalyzed reactions. One reason for the importance of this reaction was that the enzyme invertase was available in a reasonably pure form by the end of the nineteenth century when the principles of enzyme kinetics were established. In some of the early literature, this enzyme was called diastase. In fact, in some of the early literature all enzymes were called diastase. A second reason for the importance of sucrose hydrolysis in the development of enzyme kinetics was that the characteristics of acid-

3

catalyzed hydrolysis of sucrose had been well established by the latter part of the nineteenth century, and this allowed comparison of the acid-catalyzed hydrolysis with the enzyme-catalyzed reaction.

The hydrolysis of sucrose is the following reaction

$$\text{sucrose} + H_2O \underset{k_{-1}}{\overset{k_1}{\rightleftharpoons}} \text{glucose} + \text{fructose}$$

In the foregoing expression, k_1 and k_{-1} are second order rate constants, i.e. the rate of the reaction is proportional to the concentration of two reactants. If the reaction were carried out in an aqueous solution where the concentration of water would be approximately 55 M and if the concentration of sucrose were 1 M or less, the concentration of water would not change appreciably during the course of the reaction. Since the concentration of water would not change significantly even if the reaction continued to completion, one can assume $k_1' = k_1(H_2O)$, where k_1' is a pseudo-first order rate constant, the rate is proportional to the concentration of one reactant. The differential equation for the disappearance of sucrose with respect to time is

$$-\frac{d(A)}{dt} = k_1'(A) - k_{-1}(P)(Q) \tag{1.1}$$

where A = sucrose, P = glucose and Q = fructose. Throughout this book it will be assumed that the activity coefficient of any reactant is unity, thus the terms *concentration* and *activity* will be assumed to be interchangeable. If the concentration of either product were equal to zero or if k_{-1} were equal to zero, the second term on the right-hand side of eq. (1.1) would be equal to zero, and eq. 1.1 would become

$$-\frac{d(A)}{dt} = k_1'(A) \tag{1.2}$$

Equation (1.2) describes a reaction which would exhibit first order kinetic behavior. A plot of the rate of disappearance of A against the concentration of A should be a straight line which should pass through the origin with a slope equal to k_1'. This is shown in Figure 1.1.

Equation (1.2) can be rearranged and expressed in integral form.

$$\int_{A_0}^{(A)} \frac{d(A)}{(A)} = -k_1' \int_0^t dt \tag{1.3}$$

The result of the integration gives,

$$\ln(A) = -k_1' t + \ln A_0 \tag{1.4}$$

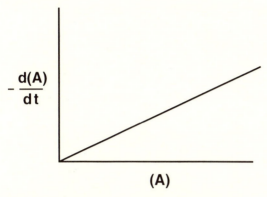

Fig. 1.1. Plot of the rate of disappearance of substrate A as a function of the concentration of A for a first order reaction.

where A_0 is the initial concentration of A.

$$(A) = A_0 e^{-k_1't} \tag{1.5}$$

Substitution of eq. (1.5) into eq. (1.2) gives,

$$-\frac{d(A)}{dt} = k_1' A_0 e^{-k_1't} \tag{1.6}$$

Plotting the rate of disappearance of A against time gives an exponential curve as is shown in Figure 1.2.

Investigations of the acid-catalyzed hydrolysis of sucrose were consistent with the hypothesis that the reaction followed first order kinetics. However,

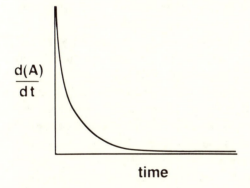

Fig. 1.2. Plot of the rate of disappearance of substrate A as a function of time for a first order reaction.

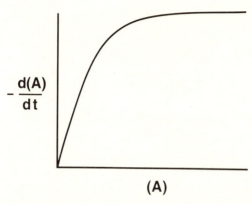

Fig. 1.3. Plot of the rate of disappearance of substrate A as a function of the concentration of A for a typical enzyme-catalyzed reaction.

investigations of the enzyme-catalyzed reaction led to observations which were perplexing at that time. Data which were obtained in experiments at low concentrations of substrate indicated first order kinetic behavior while experiments at high substrate concentrations suggested zero order kinetic behavior. That is, the reaction rate was a constant independent of substrate concentration. A careful analysis of the various results indicated that a plot of the rate of sucrose disappearance against sucrose concentration had the appearance shown in Figure 1.3. Numerous hypotheses were advanced to explain the kinetic behavior of enzyme-catalyzed processes[1], but none received widespread acceptance until the proposal suggested by Brown[2]. Brown's hypothesis was influenced by observations made by others. Wirtz[3] had reported that the proteolytic enzyme papain formed an insoluble complex with the substrate fibrin. This indicated that enzymes could combine with their substrates, but it did not provide evidence that the resulting complex was an obligatory intermediate in the reaction sequence. Additionally, O'Sullivan and Tompson[4] observed that invertase could tolerate a higher temperature in the presence of its substrate than in the absence of substrate. This observation was consistent with the hypothesis that sucrose could combine with invertase to form a complex which was more resistant to heat inactivation than was the native enzyme. Once again, this did not mandate that the enzyme-substrate complex was an obligatory intermediate in the reaction sequence. Finally, Emil Fischer's[5] "lock and key" explanation for enzyme specificity was best interpreted, at that time, in terms of an enzyme-substrate complex which is an obligatory intermediate in the reaction sequence. Thus, Brown suggested the following model for enzyme-catalyzed reactions.

$$E + A \underset{k_{-1}}{\overset{k_1}{\rightleftharpoons}} EA \overset{k_2}{\rightarrow} E + P$$

In the foregoing reaction sequence, E represents the free enzyme while EA represents the complex of enzyme with substrate A. In this model, and throughout this text, the letters A, B, C, and D represent substrates while the letters P, Q, R and S represent products of the enzymic reaction. The forgoing model predicts that the reaction rate, i.e the increase of P with time should be $v = k_2 (EA)$. Thus the rate of the reaction is proportional to the concentration of the EA complex. If the concentration of the enzyme were held constant and assays were run at increasing concentrations of substrate A, one would expect the concentration of the EA complex to be proportional to the concentration of A at low concentrations of A. Under those conditions, the kinetic behavior of the reaction would approximate first order kinetics. Inspection of Figure 1.3 shows that, at low concentrations of substrate A, the rate of the reaction is approximately a linear function of substrate concentration. On the other hand, if the concentration of the substrate were so high that essentially all the enzyme was present in the form of the EA complex, the rate of the reaction would be determined by the rate of decomposition of the EA complex to form a free enzyme and the product. At that point, increasing the concentration of the substrate would have no further effect on the rate of the reaction, and the reaction would exhibit zero order kinetic behavior with respect to substrate concentration. Inspection of Figure 1.3 shows that, at the highest concentrations of substrate, the plot of reaction rate versus the concentration of substrate is approximately a straight line with slope equal to zero. At intermediate concentrations of substrate, the curve represents a transition from first order to zero order kinetic behavior. The complete plot of reaction rate versus substrate concentration (substrate-saturation curve) is that of a rectangular hyperbola.

1.2 The basic assumptions on which derivation of an equation is based

The model which Brown proposed for enzymic reactions has withstood the test of time, but it is strictly intuitive and lacks a mathematical foundation. A mathematical treatment of this model was advanced first by a brilliant French scientist, Victor Henri[6]. It was Henri who derived the equation which is often attributed to Michaelis and Menten. Indeed, Michaelis and Menten[7] acknowledged that the purpose of their work was to provide experimental affirmation of the mathematical formulation published by Henri. Based on the model proposed by Brown, one can write differential equations for the change in concentrations of each of the two enzyme species with respect to time.

$$\frac{d(E)}{dt} = -k_1(E)(A) + (k_{-1} + k_2)(EA) \tag{1.7}$$

$$\frac{d(EA)}{dt} = k_1(E)(A) - (k_{-1} + k_2)(EA) \tag{1.8}$$

As pointed out earlier,

$$v = -\frac{d(A)}{dt} = k_2(EA) \tag{1.9}$$

One might think that a mathematical expression for the rate of an enzymic reaction could be obtained by an analytical solution of the system of differential equations expressed in eqs. (1.7) and (1.8) and substitution of the expression for (EA) into eq. (1.9). Unfortunately, there is no analytical solution of eqs. (1.7) and (1.8). However, Henri reasoned that within a few milliseconds after the mixing of the enzyme with its substrate the concentrations of free enzyme and enzyme-substrate complex would become time-invariant. That is, for a given concentration of enzyme the relative amount of free enzyme and enzyme-substrate complex would be a function of substrate concentration, but the actual amount of each enzyme species would remain constant after the first few milliseconds. This assumption allows the differential equations of eqs. (1.7) and (1.8) to be replaced by the following linear algebraic equations.

$$-k_1(E)(A) + (k_{-1} + k_2)(EA) = 0 \tag{1.10}$$

$$k_1(E)(A) - (k_{-1} + k_2)(EA) = 0 \tag{1.11}$$

The foregoing two equations contain two unknown quantities, namely (E) and (EA), but it is not possible to solve the unknown quantities because the equations are not independent; in fact for the model under consideration they are identical. In order to derive an equation for the rate of an enzyme-catalyzed reaction, it is necessary to make a number of assumptions. These are,

$$E_t = (E) + (EA) \tag{1.12}$$

$$A_t \gg E_t \tag{1.13}$$

$$\frac{d(E)}{dt} = \frac{d(EA)}{dt} = 0 \tag{1.14}$$

$$(P) = 0 \tag{1.15}$$

In eq. (1.13), A_t is the total substrate concentration. The first three of these assumptions are essential for the derivation of the rate equation, the fourth assumption is made at this point as a matter of convenience and in chapter 4 the restriction imposed by eq. (1.15) will be removed. It is imperative that the reason for and the implications and validity of the assumptions expressed in eqs. (1.12) through (1.14) be understood. The logic behind the assumption expressed in eq. (1.12) is obvious. One could not conduct a valid assay if the total activity of the enzyme were changing during the assay. This assumption is often termed the *enzyme conservation expression*. However, this equation is indispensable mathematically for it provides a third equation and, therefore, a total of two independent equations which can be solved for the two unknown quantities. The reason for the remaining assumptions will be discussed in subsequent chapters.

1.3 The Briggs-Haldane steady state treatment of enzyme kinetic behavior

The derivation which will be presented is neither that of Henri nor that of Michaelis and Menten, but rather, the derivation of Briggs and Haldane[8,9]. The reason for following the Briggs-Haldane derivation is that it is a more general treatment. As noted earlier, the rate of the enzyme-catalyzed reaction for the model under consideration is $v = k_2(EA)$. The concentration of the free enzyme can be obtained from either eq. (1.10) or eq. (1.11).

$$(E) = \frac{(k_{-1} + k_2)}{k_1(A)}(EA) \tag{1.16}$$

Equation (1.16) can be substituted into eq. (1.12) and rearranged as

$$(EA) = \frac{k_1 E_t(A)}{k_{-1} + k_2 + k_1(A)} \tag{1.17}$$

The rate of the reaction is obtained by multiplying eq. (1.17) by k_2.

$$v = \frac{k_1 k_2 E_t(A)}{k_{-1} + k_2 + k_1(A)} \tag{1.18}$$

Equation (1.18) is identified as the Briggs-Haldane equation, the Michaelis-Menten equation and the Henri equation. Traditionally it is called the Michaelis equation and, reluctantly, that tradition will be followed in this book.

Equation (1.18) expresses the rate equation in terms of rate constants for the individual reactions. Throughout this book an enzymic rate equation expressed in terms of rate constants will be called the *rate equation* in the coefficient form. While the rate equation is usually derived in this form, it is not a useful form of the rate equation because most of the rate constants are generally inaccessible in investigations of the steady state behavior of enzymes. For this reason it is necessary to reformulate eq. (1.18) such that it is expressed in terms of parameters which can be determined in steady state studies. Throughout this book, these reformulations will be conducted in a similar manner. Equation (1.18) can be re-written as

$$v = \frac{\text{num. } 1(A)}{\text{constant} + \text{coef. } A(A)},$$ (1.19)

where num. $1 = k_1 k_2 E_t$, constant $= k_{-1} + k_2$, and coef. $A = k_1$. The equation is reformulated by dividing both the numerator and denominator of the right hand side of eq. (1.19) by coef. A. The result is,

$$v = \frac{\dfrac{\text{num. } 1}{\text{coef. } A}(A)}{\dfrac{\text{constant}}{\text{coef. } A} + (A)}$$ (1.20)

The coefficient of the numerator term in eq. (1.20) is a constant, and the first term in the denominator of eq. (1.20) is also a constant. The equation is reformulated as,

$$v = \frac{V_{\max}(A)}{K_m + (A)},$$ (1.21)

where $V_{\max} = k_2 E_t$ and $K_m = (k_{-1} + k_2)/k_1$. The Michaelis constant is K_m, and $V_{\max}$ is the maximal velocity. More precisely, $V_{\max}$ is the velocity of the reaction when the enzyme is saturated with the substrate. Throughout this book an enzymic rate equation expressed in terms of the steady state parameters will be called a *rate equation* in the kinetic form. Later in this chapter methods which provide for estimation of the steady state parameters, $V_{\max}$ and K_m will be discussed. Equations (1.18) and (1.21) both describe a rectangular hyperbola. Stated in more descriptive terms, they are 1:1 order rational polynomials. A rational polynomial is a ratio of polynomials. A 1:1 order rational polynomial contains the independent variable, (A) in this case, to the first power in both the numerator and denominator.

It is informative to divide both the numerator and denominator of the right hand side of eq. (1.21) by the concentration of A,

$$v = \frac{V_{max}}{1 + \frac{K_m}{(A)}} \tag{1.22}$$

If the concentration of A were much less than K_m, such that $1 \ll K_m/(A)$, eq. (1.22) would become

$$v = \frac{V_{max}}{K_m}(A). \tag{1.23}$$

This equation describes a reaction which exhibits first order kinetic behavior, and the apparent first order rate constant is V_{max}/K_m. However, if the concentration of substrate were so great that $K_m/(A) \simeq 0$, eq. (1.22) would become

$$v \simeq V_{max}. \tag{1.24}$$

At this point the rate of the reaction would be independent of (A) and the reaction would exhibit zero order kinetic behavior with respect to substrate concentration. The mathematical definition of saturation of the enzyme with substrate A is, $K_m/(A) \simeq 0$. Finally, if $K_m = (A), v = 1/2(V_{max})$. Thus the K_m is the concentration of the substrate which results in half maximal velocity, and the Michaelis constant is expressed in molarity.

The derivation presented here is that of Briggs and Haldane, and it differs from that developed by Henri and also that employed by Michaelis and Menten. In the case of the model under consideration, k_2 is the rate constant which includes the step which usually involves either the cleavage or formation of a covalent bond. If this step were very much slower than the other steps in the model, an equilibrium would be established between the free enzyme and the substrate and the enzyme-substrate complex. If such were the case, k_2 would be much smaller than k_{-1} and the Michaelis constant would be $K_m \simeq k_{-1}/k_1$. Thus, in the Briggs-Haldane treatment the Michaelis constant is a kinetic constant while in the Henri treatment it is a dissociation constant and, therefore, a thermodynamic constant. The matter of whether or not the reaction involving the cleavage of formation of a covalent bond is very much slower than other steps in the reaction sequence will be discussed in chapter 9 of this book.

1.4 Estimation of steady state parameters

If one were to measure the velocity of an enzyme-catalyzed reaction in a series of assays in which the substrate concentration in each assay varied from one which was sufficiently small to result in a low rate relative to V_{max} to one where the substrate concentration were large enough to result in maximal velocity, one could plot the data and estimate both V_{max} and K_m. However, if, for example, the solubility of the substrate were limited in an aqueous solution, it might be impossible to estimate V_{max} and therefore K_m could also not be estimated. For this reason, efforts were made to rearrange eq. (1.18) in a linear form so that V_{max} could be obtained by extrapolation. Haldane and Stern[10], following the suggestions of B. Woolf, rearranged eq. (1.21) by dividing both sides of the equation by (A) and then inverting both sides of the equation to obtain

$$\frac{(A)}{v} = \frac{1}{V_{max}}(A) + \frac{K_m}{V_{max}}. \tag{1.25}$$

Equation (1.25) describes a linear relationship if $(A)/v$ were plotted against (A). The slope of the line is the reciprocal of V_{max} and the intercept of the $(A)/v$ axis is K_m/V_{max}. This same rearrangement was proposed by Hanes[11]. Haldane and Stern also noted that multiplying both sides of eq. (1.21) by $[K_m + (A)]$ and the rearrangement gives,

$$v = -K_m\frac{v}{(A)} + V_{max}. \tag{1.26}$$

Equation (1.26) is a linear relationship whose slope is $-K_m$ and whose intercept of the v axis is V_{max}. Lineweaver and Burk[12] utilized yet another rearrangement to obtain a linear form of eq. (1.21). This was accomplished by simply inverting both sides of eq. (1.21).

$$\frac{1}{v} = \frac{K_m}{V_{max}}\frac{1}{(A)} + \frac{1}{V_{max}}. \tag{1.27}$$

The resulting equation is that of a straight line whose slope is K_m/V_{max} and whose intercept of the $1/v$ axis is $1/V_{max}$. The point of intersection of the $1/(A)$ axis is $-1/K_m$.

There are additional linear forms of eq. (1.21), for example, the direct linear plot of Eisenthal and Cornish-Bowden[13], but the foregoing are the most widely used. It is important to recognize that these equations do not give rise to estimates of the steady state kinetic parameters with equal degrees of precision. Note that $1/v$, the dependent variable in the Lineweaver-Burk equation [eq. (1.27)] approaches infinity as $1/(A)$, the independent variable, approaches infinity. Hence, the Lineweaver-Burk plot places maximum weight on those observations which are made at low concentrations of the substrate, and those are the values which are associated normally with the largest experimental error. The converse is true of eq. (1.25) while eq. (1.26) places uniform weight on observations throughout the substrate-saturation curve. The significance is that the Lineweaver-Burk plot is the least desirable method for obtaining quantitative estimates of the steady state parameters. The objection to the use of the Lineweaver-Burk plot can be alleviated to some degree by utilizing a statistical program which employs weighting factors, but this necessitates the selection of an appropriate weighting factor. It is important to realize that the foregoing objection applies to the use of the Lineweaver-Burk plot as a means of obtaining quantitative estimates of K_m and V_{max} only, it does not argue against plotting data as a double reciprocal plot utilizing the estimates of the parameters which have been obtained by a more satisfactory method. Probably the most feasible method of obtaining quantitative estimates of the steady state parameters is the nonlinear regression method of Wilkinson[14]. The original publication outlines the method clearly for use with a calculator, but the procedure was outlined so well in Wilkinson's publication that it is easy to adapt it to a computer program to be run on a personal computer or even a programmable hand held calculator. This procedure is so elegant that there is little reason to obtain estimates of V_{max} and K_m by any other method provided the substrate-saturation curve is a rectangular hyperbola and the procedure is outlined in section 1.A.3 of the appendix to this chapter.

1.5 Problems for chapter 1

1.1 Derive expressions for the fraction of the total enzyme present as the free enzyme and for the fraction of the total enzyme present as the EA complex for the enzyme model considered in this chapter.

1.2 On the same sheet of graph paper, plot the $(E)/E_t$ and $(EA)/E_t$ as a function of $(A)/K_m$. Vary $(A)/K_m = 0.1$ to 10.

1.3 The following data were obtained in a substrate-saturation experiment.

(A)	v
mM	μmoles/minute
0.075	0.0120
0.100	0.0152
0.150	0.0205
0.200	0.0245
0.250	0.0280

Estimate K_m and V_{max} from a plot of $1/v$ versus $1/(A)$, and from a plot of $(A)/v$ versus (A), and finally from a plot of v versus $v/(A)$.

Appendix: A brief look at statistical analysis

1.A.1 Definition of a few statistical terms

It is the purpose of this appendix to provide a brief account of the simpler statistical analyses employed in enzyme kinetics. The first statistic is the arithmetic mean or average. If one were to measure the change in absorbance at 340 mμ in a cuvette in a given time interval after a dehydrogenase had been added to a reaction medium containing NAD^+ and the appropriate oxidizable substrate several times, one would record a number of slightly different values. If the several values were designated $\sum Y_i$ and n were the number of observations, the average change in absorbance would be

$$\bar{Y} = \frac{\sum Y_i}{n} \tag{1.A.1}$$

The arithmetic mean does not give any indication of the amount of scatter in the observations. A measure of the accuracy of the mean should be related to the deviations about the mean, but, in theory, the sum of the deviations greater than the mean should be offset by the deviations less than the mean. Hence the sum of the deviations should be equal to zero. For this reason, and for theoretical reasons that will not be discussed here[15], the deviations about the mean are squared. The variance is defined as the sum of the squares of the deviations divided by the degrees of freedom. If there is one parameter measured, the degrees of freedom is given by $n - 1$. Therefore the expression for the variance is,

$$\text{variance} = s^2 = \frac{\sum (Y - \bar{Y})^2}{n - 1} \tag{1.A.2}$$

The squared term in the numerator can be expanded as

$$\sum (Y^2 - 2Y\bar{Y} + \bar{Y}^2) = \sum Y^2 - 2\sum Y \sum \frac{Y}{n} + \left(\frac{\sum Y}{n}\right)^2 n$$

The variance can be expressed as follows:

$$s^2 = \frac{\sum Y^2 - \dfrac{(\sum Y)^2}{n}}{n - 1} \tag{1.A.3}$$

The standard deviation is defined as the square root of the variance.

$$\text{s.d.} = \sqrt{\frac{\sum Y^2 - \frac{(\sum Y)^2}{n}}{n-1}} \qquad (1.A.4)$$

Standard deviation and standard error are similar terms. They may be used interchangeably if the statistic to which they apply is specified. The coefficient of variation is defined as follows:

$$\text{c.v.} = \frac{\sqrt{\text{variance}}}{\overline{Y}} \times 100 = \frac{S}{\overline{Y}} \times 100 \qquad (1.A.5)$$

The symbol s^2 is defined as the variance of a given sample while σ^2 is defined as the variance of the population from which the sample is taken. The former is estimated from the data, but the statistician is usually interested in the variance of the population rather than that of the sample of the population. In the same manner, the arithmetic mean of the sample is defined as $\overline{Y}$ while the arithmetic mean of the population is μ.

1.A.2 Linear regression

A procedure which is employed extensively in the analysis of enzyme kinetic data, as well as in all of biochemistry, is linear regression. Linear relationships are well understood in mathematical terms. Three linear transformations of the Michaelis equation have been presented in this chapter. In general terms, the assumption of a linear relationship implies that a dependent variable, Y, is a linear function of an independent variable X. In the case of the Lineweaver-Burk plot, $1/v$ is the dependent variable while $1/(A)$ is the independent variable. In the case of the $(A)/v$ versus (A) plot, $(A)/v$ is the dependent and (A) is the independent variable. However, the linear regression, as here presented, is based on four assumptions. These assumptions are,

1) It is assumed that the independent variable, X, is measured without error.
2) The expected value of the dependent variable for a given value of the independent variable is,

$$Y = A + BX. \qquad (1.A.6)$$

3) For any given value of X, the observed Y values are distributed independently and normally. This is represented by,

$$Y_i = A + BX + \varepsilon_i. \qquad (1.A.7)$$

where ε_i is the error in the estimate.
4) It is assumed that the variance around the regression line is constant and, therefore, independent of the magnitude of X or Y.

The aim in the linear regression is to calculate the values of A and B in eq. (1.A.7) such that Y can be estimated for any given value of X. Thus, the difference between the observed values of Y_i and the value estimated from the regression line would be

$$r_i = \hat{Y}_i - Y_i \qquad (1.A.8)$$

where r_i is the residual and $\hat{Y}_i$ is the estimated value of Y_i. As with variance, the residuals are squared.

$$\sum r_i^2 = \sum (\hat{Y}_i - Y_i)^2 = \sum (A + BX_i - Y_i)^2 \tag{1.A.9}$$

The problem is to find the values for A and B which minimize $\sum r_i^2$. This is accomplished by setting the following derivatives equal to zero.

$$\frac{d\sum (A + BX - Y)^2}{dA} = 0, \quad \frac{d\sum (A + BX - Y)^2}{dB} = 0$$

An expansion of the squared numerator and differentiation gives the following expressions.

$$\frac{d\sum r^2}{dA} = 2\sum A + 2B\sum X - 2\sum Y = 0$$

$$\frac{d\sum r^2}{dB} = 2A\sum X + 2B\sum X^2 - 2\sum XY = 0$$

The following equations arise from the foregoing

$$\sum Y = nA + B\sum X \tag{1.A.10}$$

$$\sum XY = A\sum X + B\sum X^2 \tag{1.A.11}$$

Equations (1.A.10) and (1.A.11) can be solved for A and B. Note that A is the point of intersection of the regression line with the Y axis and B is the slope of the regression line, but the latter is more often referred to as the *regression coefficient*.

A variety of computer-based library programs which provide for numerical solution of eqs. (1.A.10) and (1.A.11) as well as providing estimates of the standard error for both parameters are available. There are also books which contain the source code for such programs[16]. Nevertheless, it seems appropriate to provide a brief account here of how such computations are conducted[17]. Equations (1.A.10) and (1.A.11) can be generalized as,

$$a_{11}x_1 + a_{12}x_2 = b_1$$

$$a_{21}x_1 + a_{22}x_2 = b_2$$

These equations can be written in matrix form

$$\begin{vmatrix} a_{11} & a_{12} \\ a_{12} & a_{22} \end{vmatrix} \begin{vmatrix} x_1 \\ x_2 \end{vmatrix} = \begin{vmatrix} b_1 \\ b_2 \end{vmatrix}$$

An augmented matrix which consists of the A matrix, the B vector from the foregoing and an identity matrix can be constructed

$$|A \quad B \quad I| = \begin{vmatrix} a_{11} & a_{12} & b_1 & 1 & 0 \\ a_{21} & a_{22} & b_2 & 0 & 1 \end{vmatrix}$$

After the equations have been solved, the resultant augmented matrix will be

$$|I \quad x \quad A^{-1}| = \begin{vmatrix} 1 & 0 & x_1 & c_{11} & c_{12} \\ 0 & 1 & x_2 & c_{21} & c_{22} \end{vmatrix}$$

In the foregoing, the elements c_{ij} are the elements in matrix A^{-1}. The inversion of matrix A is accomplished by two types of operations. The first of these is *normalization* in which all of the elements in a row of the original augmented matrix are divided by the first non-zero element of the row. By a repetition of this operation, the diagonal elements of A are converted to ones. The second operation is *reduction* in which the non-diagonal elements of A are converted to zeros. Normalization of the first row of the original augmented matrix gives

$$\begin{vmatrix} 1 & \dfrac{a_{12}}{a_{11}} & \dfrac{b_1}{a_{11}} & \dfrac{1}{a_{11}} & 0 \\[2ex] a_{21} & a_{22} & b_2 & 0 & 1 \end{vmatrix}$$

Reduction is performed by multiplying each element of the normalized first row by a_{21} and subtracting the product from the corresponding element in row 2. The result of this reduction is

$$\begin{vmatrix} 1 & \dfrac{a_{12}}{a_{11}} & \dfrac{b_1}{a_{11}} & \dfrac{1}{a_{11}} & 0 \\[3ex] 0 & \dfrac{a_{11}a_{22}-a_{12}a_{21}}{a_{11}} & -\dfrac{a_{11}b_2-a_{21}b_1}{a_{11}} & -\dfrac{a_{21}}{a_{11}} & 1 \end{vmatrix}$$

Normalization of the second row gives

$$\begin{vmatrix} 1 & \dfrac{a_{12}}{a_{11}} & \dfrac{b_1}{a_{11}} & \dfrac{1}{a_{11}} & 0 \\[3ex] 0 & 1 & \dfrac{a_{11}b_2-a_{21}b_1}{a_{11}a_{22}-a_{12}a_{21}} & -\dfrac{a_{21}}{a_{11}a_{22}-a_{12}a_{21}} & \dfrac{a_{11}}{a_{11}a_{22}-a_{12}a_{21}} \end{vmatrix}$$

The final reduction is accomplished by multiplying each element of the normalized second row by a_{12}/a_{11} and subtracting the product from the corresponding element of row 1. The result is the final augmented matrix.

$$\begin{vmatrix} 1 & 0 & \dfrac{a_{22}b_1-a_{21}b_2}{a_{11}a_{22}-a_{12}a_{21}} & \dfrac{a_{22}}{a_{11}a_{22}-a_{12}a_{21}} & -\dfrac{a_{21}}{a_{11}a_{22}-a_{12}a_{21}} \\[3ex] 0 & 1 & \dfrac{a_{11}b_2-a_{21}b_1}{a_{11}a_{22}-a_{12}a_{21}} & -\dfrac{a_{21}}{a_{11}a_{22}-a_{12}a_{21}} & \dfrac{a_{11}}{a_{11}a_{22}-a_{12}a_{21}} \end{vmatrix}$$

The symbols from eqs. (1.A.10) and (1.A.11) can be substituted into the foregoing matrix to give

$$\begin{vmatrix} 1 & 0 & \dfrac{\sum X^2 \sum Y - \sum X \sum XY}{n\sum X^2 - (\sum X)^2} & \dfrac{\sum X^2}{n\sum X^2 - (\sum X)^2} & -\dfrac{\sum X}{n\sum X^2 - (\sum X)^2} \\[3ex] 0 & 1 & \dfrac{n\sum XY - \sum X \sum Y}{n\sum X^2 - (\sum X)^2} & -\dfrac{\sum X}{n\sum X^2 - (\sum X)^2} & \dfrac{n}{n\sum X^2 - (\sum X)^2} \end{vmatrix}$$

It can be seen from the foregoing that A in eqs. (1.A.10) and (1.A.11) is equal to the first element in the third column of resultant augmented matrix, and B is equal to the second element in the third column of the matrix. Thus,

$$A = \frac{\sum X^2 \sum Y - \sum X \sum XY}{n \sum X^2 - (\sum X)^2} = \frac{(\sum X^2 \sum Y - \sum X \sum XY)/n}{\sum X^2 - \frac{(\sum X)^2}{n}} \qquad (1.A.12)$$

$$B = \frac{n \sum XY - \sum X \sum Y}{n \sum X^2 - (\sum X)^2} = \frac{\sum XY - \sum X \frac{\sum Y}{n}}{\sum X^2 - \frac{(\sum X)^2}{n}} \qquad (1.A.13)$$

The variance about the regression line is estimated by

$$s^2 = \frac{\sum (y - A - BX)^2}{n - 2} = \frac{\sum Y^2 - A \sum Y - B \sum XY}{n - 2} \qquad (1.A.14)$$

The degree of freedom in eq. (1.A.14) is $n - 2$ because both A and B are estimated from the observed data. Without providing a derivation, the standard error on the estimate of A is

$$\text{s.e.}_A = \sqrt{s^2 \times \frac{\sum X^2}{n \sum X^2 - (\sum X)^2}} \qquad (1.A.15)$$

It should be noted that the standard error for A is equal to the square root of the variance times the first element of the fourth column of the augmented matrix. In like manner, the standard error for the estimation of B, the regression coefficient, is

$$\text{s.e.}_B = \sqrt{s^2 \times \frac{n}{n \sum X^2 - (\sum X)^2}} \qquad (1.A.16)$$

The standard error of the regression coefficient is equal to the square root of the variance times the second element of the last column in the augmented matrix. Thus the diagonal elements in the inverse matrix A^{-1} are factors used in estimation of the standard errors. The variables which are required for estimation of A and B and their standard errors are $n, \sum Y, \sum Y^2, \sum X, \sum X^2, \sum XY$. The equations presented in this text may appear to differ from those found in reference books on statistical analysis[15], but they are equivalent. Statisticians rearrange the equations to improve the computational efficiency. The purpose here has been to outline the derivations in a fairly straightforward manner.

1.A.3 Non-linear regression in enzyme kinetic analysis

The non-linear regression method proposed by Wilkinson for estimation of the steady state enzyme kinetic parameters will be outlined here in essentially the manner it has been presented in the original publication[14]. Non-linear regression requires a preliminary estimate of the parameters, and these are obtained by linear regression similar to that described in the previous discussion. However, Wilkinson employed a weighted linear regression of (A)/v versus (A). The following expressions provide for the preliminary estimates of the parameters.

$$|D| = \sum v^3 \sum \frac{v^4}{(A)^2} - \sum \frac{v^3}{(A)} \sum \frac{v^4}{(A)} \tag{1.A.17}$$

$$K_m = \left[\sum v^4 \sum \frac{v^3}{(A)} - \sum v^3 \sum \frac{v^4}{(A)} \right] \Big/ |D| \tag{1.A.18}$$

$$V_{max} = \left[\sum v^4 \sum \frac{v^4}{(A)^2} - \left(\sum \frac{v^4}{(A)} \right)^2 \right] \Big/ |D| \tag{1.A.19}$$

The non-linear aspect of the Wilkinson method is based on the assumption that if a function is non-linear in a parameter, c, the following linear approximation may be used.

$$f_{v,c} \cong f_{v,c^0} + (c - c^0) f'_{v,c^0} \tag{1.A.20}$$

where c^0 is a provisional estimate of c and f' is the first derivative of f with respect to c. In terms of enzyme kinetics, eq. (1.A.20) becomes

$$V \cong \frac{V_{max}}{V_{max}^0} \left[\frac{V_{max}^0 (A)}{K_m^0 + (A)} - (K_m - K_m^0) \frac{V_{max}^0}{(K_m^0 + (A))^2} \right] \tag{1.A.21}$$

To initiate the calculations, the preliminary estimates of K_m and V_{max} from linear regression are used as the provisional estimates of the parameters. The following calculations lead to updated estimates of the parameters.

$$f = \frac{V_{max}^0 (A)}{K_m^0 + (A)} \tag{1.A.22}$$

$$f' = -\frac{V_{max}^0 (A)}{(K_m^0 + (A))^2} \tag{1.A.23}$$

$$|D| = \sum f^2 \sum f'^2 - (\sum ff')^2 \tag{1.A.24}$$

$$b_1 = [\sum f'^2 2 \sum vf - \sum ff'^2 \sum vf'] / |D| \tag{1.A.25}$$

$$b_2 = [\sum f^2 \sum vf' - \sum ff' \sum vf] / |D| \tag{1.A.26}$$

$$K_m = K_m^0 + \frac{b_2}{b_1} \tag{1.A.27}$$

$$V_{max} = V_{max}^0 \times b_1 \tag{1.A.28}$$

The updated estimate of K_m is tested against the provisional estimate. If the two are sufficiently close, for example, if $abs(K_m - K_m^0)/K_m^0 \lesssim 0.001$, the updated parameters are accepted as the best estimates. If on the other hand, the updated K_m does not pass the test, the provisional estimates of the parameters are replaced by the updated estimates and the calculations embodied in eqs. (1.A.22) through (1.A.28) are repeated. It should be understood that, as K_m^0 approaches the best estimate, the value of b_1 approaches unity and b_2 approaches zero. After the best estimates have been obtained, the variance can be calculated as

$$s^2 = \frac{\sum v^2 - b_1 \sum vf - b_2 \sum vf'}{n - 2} \tag{1.A.29}$$

The standard error for K_m is

$$\text{s.e.}_{K_m} = \frac{1}{b_1}\sqrt{s^2 \times \frac{\sum f^2}{|D|}} \tag{1.A.30}$$

while the standard error for V_{max} is

$$\text{s.e.}_{V_{max}} = V_{max}\sqrt{s^2 \times \frac{\sum f'^2}{|D|}} \tag{1.A.31}$$

A word of caution should be introduced at this point. The Wilkinson procedure is based on the assumption that eq. (1.21) describes the kinetic behavior of the enzyme accurately. In some of the subsequent chapters of this book, enzyme models which give rise to rate equations which are not consistent with eq. (1.21) will be considered. The Wilkinson procedure cannot be employed to obtain meaningful estimates of parameters in the case of those enzymes. The Wilkinson procedure converges rapidly on the best estimate of K_m if eq. (1.21) is appropriate. If a computer program is written to execute the Wilkinson procedure, it is recommended that a counter be included to record the total iterations required for convergence. If more than 3 or 4 iterations are required, one would be well advised to question whether eq. (1.21) is appropriate for the enzyme involved.

References

1. Segal, H. L. (1959). in *The Enzymes*, Vol. 1, 2d. Ed. (P. D. Boyer, H. Lardy and K. Myrback, Ed.), pp. 1–48, New York, Acad. Press.
2. Brown, A. J. (1902). Enzyme action, *J. Chem. Soc.* **81**: 373–88.
3. Wirtz, A. (1880). Compt. rend. **91**: 787.
4. O'Sullivan, F. R. S. and Tompson, F. W. (1890). *J. Chem. Soc.* **57**: 834.
5. Fischer, E. (1894). Einfluss der configuration auf die wirkung der enzyme. *Dtsh. Chem. Ges. Ber.* **27**: 2985–93.
6. Henri, V. (1903). *Lois generales de l'action des diastases*, Paris, Hermann.
7. Michaelis, L. and Menten, M. L. (1913). Die kinetik der invertin wirkung. *Biochem. Z.* **49**: 333–69.
8. Briggs, G. E. and Haldane, J. B. S. (1925). A note on the kinetics of enzyme action. *Biochem. J.* **19**: 338–39.
9, Haldane, J. B. S. (1930). *Enzymes*, pp. 28–64, London, Longmans, Green and Co.
10. Haldane, J. B. S. and Stern, K. (1932). in *Wissenschafliche Forschunsberichte Naturwissenshaftleche*, Band 28, (R. E. Liesegung, Ed.), Theo. Steinkopf, Dresden.
11. Hanes, C. S. (1932). The effect of starch concentration upon the velocity of hydrolysis by the amylase of germinated barley. *Biochem. J.* **26**: 1406–21.
12. Lineweaver, H. and Burk, J. (1934). The determination of enzyme dissociation constants. *J. Am. Chem. Soc.* **56**: 658–66.
13. Eisenthal, R. and Cornish-Bowden, A. (1974). The direct linear plot. *Biochem. J.* **139**: 715–20.
14. Wilkinson, G. (1961). Statistical estimations in enzyme kinetics. *Biochem. J.* **80**: 324–332.

15. Sokal, R. E. and Rohlf, F. J. (1981). *Biometery*, 2d. Ed., pp. 454–560, San Francisco, Freeman & Co.
16. Carnahan, B., Luther, H. A. and Wilkes, J. O. (1969). *Applied Numerical Methods*, pp. 571–73, New York, Wiley.
17. Carnahan, B., Luther, H. A. and Wilkes, J. O. (1969). *Applied Numerical Methods*, 269–73, New York, Wiley.

2
A closer look at the basic assumptions

The derivation of the Michaelis equation in the previous chapter was based on four assumptions. It was stated that an understanding of these assumptions is essential to an understanding of steady state enzyme kinetics. The requirement for the first assumption, eq. (1.12), was discussed in chapter 1. In this chapter the significance of the second and third assumptions will be investigated.

2.1 Why must the substrate concentration greatly exceed that of the enzyme?

The second assumption is given in eq. (1.13), but it is re-stated here, $A_t \gg E_t$. Of the four assumptions, this one is most widely misinterpreted. The presumption is *not* that the enzyme must be saturated with substrate! The purpose of this assumption is to guarantee that there is not a significant fraction of the substrate bound to the enzyme during the assay. Equation (1.12) is an enzyme conservation equation, but a conservation equation was not included for the substrate. A substrate conservation equation for the model considered in chapter 1 is

$$A_t = (A) + (EA) + (P). \tag{2.1}$$

The implicit assumption in the derivation presented in chapter 1 is that $A_t = (A)$. Equation (1.15) stipulates that $(P) = 0$, but even so, A_t, the total substrate, will be equal to the free substrate, (A), only if (EA) does not represent a significant fraction of the total substrate. In the case of most enzymes *in vitro*, this condition is satisfied easily. The following calculation illustrates this contention. Suppose the enzyme being studied has a turn-over number of 10,000, that is, 1 μmole of enzyme, when saturated with the substrate, would catalyze the conversion of 10,000 μmoles of substrate to product per minute. This is actually a low turnover number. Furthermore, suppose the substrate concentration were equal to one tenth of the Michaelis constant and that sufficient enzyme were present in the assay medium to catalyze the turnover of 1 percent of the substrate per minute. When these values are substituted into eq. (1.21), one has

$$0.01 \,\mu\text{moles minutes}^{-1} \times A_t = \frac{10,000 \,\mu\text{moles minute}^{-1} \times 0.1 \, K_m \times E_t}{1.1 \times K_m} \tag{2.2}$$

In this example $A_t/E_t = 90,000$, so the assumption would be satisfied easily.

2.2 What if the substrate concentration does not greatly exceed that of the enzyme?

If the substrate of the enzymic reaction were a large molecule, such as is the case for protein kinases or protein phosphatases, the assumption might be difficult to achieve. In such cases, a substrate conservation equation must be included in the derivation of the rate equation. Substitution of a substrate conservation expression into the equation for the concentration of (EA), eq. (1.17) gives,

$$(EA) = \frac{E_t[A_t - (EA)]}{K_m + A_t - (EA)}. \tag{2.3}$$

Rearrangement of eq. (2.3) results in the following polynomial,

$$(EA)^2 - (K_m + A_t + E_t)(EA) + E_t A_t = 0. \tag{2.4}$$

Equation (2.4) can be solved for the concentration of the EA complex and multiplication by k_2 gives the equation for the rate of the reaction.

$$v = \frac{k_2}{2}[(K_m + A_t + E_t) \pm \sqrt{(K_m + A_t + E_t)^2 - 4E_t A_t}] \tag{2.5}$$

Equation (2.5) can be simplified somewhat by recognizing that the velocity of the enzyme-catalyzed reaction must be equal to zero if the concentration of either the substrate or the enzyme were equal to zero. Inspection of eq. (2.5) shows that the velocity would be equal to zero only in the case of the negative sign before the radical sign.

$$v = \frac{k_2}{2}[(K_m + A_t + E_t) - \sqrt{(K_m + A_t + E_t)^2 - 4E_t A_t}] \tag{2.6}$$

However, even with this simplification, eq. (2.6) differs sufficiently from the form of the Michaelis equation that one might well question its validity.

Since the second term under the radical sign is negative, one might question if the conditions might exist where the value of the radicand could be negative. If such were the case the equation would give rise to an imaginary root. Reiner examined this possibility in his excellent book[1]. The expansion of the terms in the radicand gives

$$K_m^2 + 2K_m A_t + 2K_m E_t + A_t^2 - 2E_t A_t + E_t^2$$

This would be equal to $(K_m + A_t - E_t)$ if the expression contained $-2K_m E_t$ rather than $2K_m E_t$. This can be accomplished by adding both $-4K_m E_t$ and $+4K_m E_t$ to the foregoing expansion. The result is $(K_m + A_t - E_t)^2 + 4K_m E_t$. Since the first term in the resulting expression is squared, it is always positive, and the second term is also positive. Therefore, eq. (2.6) cannot give rise to an imaginary root. Equation (2.6) should be subjected to one additional test. If the substrate concentration were increased until it became infinitely large, eq. (2.6) should give the same result as eq. (1.18). This can be investigated by rearranging eq. (2.6).

$$v = \frac{k_2}{2}\left[(K_m + A_t + E_t) - (K_m + A_t + E_t)\sqrt{1 - \frac{4E_t A_t}{(K_m + A_t + E_t)^2}}\,\right] \tag{2.7}$$

The radicand can be expanded by the binomial theorem for fractional powers as $\sqrt{1 + x} = 1 + (1/2)x +$ higher powers of x. Since the second term under the radical sign in eq. (2.7) approaches zero as A_t approaches infinity, the terms in higher powers of x can be ignored. Under these conditions, eq. (2.7) becomes

$$v = \simeq \frac{k_2}{2}\left[A_t - A_t\left(1 - \frac{2E_t A_t}{A_t^2}\right)\right] = k_2 E_t \tag{2.8}$$

Thus, eq. (2.8) predicts that the rate of the reaction will equal V_{max} at high concentrations of the substrate. Equation (2.6) is an appropriate expression for the rate of the reaction when the assumption $A_t \gg E_t$ is not satisfied, but it is not a convenient equation with which to work.

If a valid estimate of maximal velocity can be obtained, an equation for the Michaelis constant can be derived by a somewhat different algebraic manipulation. Substitution of the substrate conservation equation into the rate equation in the kinetic form, eq. (1.21), gives,

$$v = \frac{V_{max}[A_t - (EA)]}{K_m + A_t - (EA)} \tag{2.9}$$

Equation (2.9) can be rearranged,

$$K_m = \left(\frac{V_{max} - v}{v}\right)A_t - \left(\frac{V_{max} - v}{v}\right)(EA) \tag{2.10}$$

Substitution of eq. (1.17) in the kinetic form into eq. (2.10) for (EA) and substitution of eq. (1.21) for v in the denominator of the second term on the right hand side of eq. (2.10) gives,

$$K_m = \left(\frac{V_{max} - v}{v}\right)A_T - \left(\frac{V_{max} - v}{V_{max}}\right)E_T. \tag{2.11}$$

An estimate of the Michaelis constant can be obtained from eq. (2.11) only if an appropriate estimate of maximal velocity is available. Otherwise, it is necessary to use the rather unwieldy eq. (2.6) if the assumption A_t/E_t cannot be satisfied.

2.3 Examination of the entire time-course of an enzymic reaction

It was stated in chapter 1 that it is not possible to obtain analytical solutions to the differential equations for the various enzyme species in an enzyme model. It is possible to obtain an analytical solution to the *approximate* differential equation for the enzyme-substrate complex for the model considered in chapter 1. The differential is approximate because it will be assumed that the concentration of substrate is constant and the concentration of the product will be assumed to be equal to zero. The approximate equation is,

$$\frac{d(EA)}{dt} = k_1[E_t - (EA)]A_t - (k_{-1} + k_2)(EA). \tag{2.12}$$

It is important to note that, in the derivation that follows, the assumptions expressed in eqs. (1.12), (1.13) and (1.15) are presumed to be satisfied. That is, only the steady state assumption, eq. (1.14) is ignored. Rearrangement of eq. (2.12) gives,

$$\frac{d(EA)}{dt} + (k_{-1} + k_2 + k_1 A_t) = k_1 E_t A_t \tag{2.13}$$

Equation (2.13) is a homogeneous linear differential equation, and its solution is given by,

$$(EA) = e^{-\int(k_{-1} + k_2 + k_1 A_t)dt}\left[k_1 E_t A_t \int e^{-\int(k_{-1} + k_2 + k_1 A_t)dt} dt + C\right] \tag{2.14}$$

$$(EA) = \frac{k_1 E_t A_t}{k_{-1} + k_2 + k_1 A_t} + C e^{-(k_{-1} + k_2 + k_1 A_t)t} \tag{2.15}$$

In the foregoing equations, C is an integration constant. The boundary conditions are that at $t = 0$, $(EA) = 0$, and thus,

$$C = -\frac{k_1 E_t A_t}{k_{-1} + k_2 + k_1 A_t} \tag{2.16}$$

Therefore,

$$(EA) = \frac{k_1 E_t A_t}{k_{-1} + k_2 + k_1 A_t}(1 - e^{-(k_{-1} + k_2 + k_1 A_t)t}) \tag{2.17}$$

The equation for the rate of the reaction is obtained by multiplying eq. (2.17) by k_2. The rate of the reaction is equal to $d(P)/dt$, and the expression can be integrated to obtain an expression for the concentration of P as a function of time. That is, by carrying out this integration an equation is derived which describes the time-course curve of the reaction.

$$\int_0^{(P)} d(P) = \frac{k_1 k_2 E_t A_t}{k_{-1} + k_2 + k_1 A_t} \int_0^t (1 - e^{-(k_{-1} + k_2 + k_1 A_t)t}) \, dt \tag{2.18}$$

$$(P) = \frac{(k_1 k_2 E_t A_t)}{k_{-1} + k_2 + k_1 A_t} + \frac{k_1 k_2 E_t A_t}{(k_{-1} + k_2 + k_1 A_t)^2}(e^{-(k_{-1} + k_2 + k_1 A_t)t} - 1) \tag{2.19}$$

The exponential term in eq. (2.19) can be expanded as the following series.

$$e^x = \sum_{r=0}^{\infty} \frac{x^r}{r!} = 1 + x + \frac{1}{2}x^2 + \frac{1}{6}x^3 + \cdots \tag{2.20}$$

The exponent x in eq. (2.20) is equal to $-(k_{-1} + k_2 + k_1 A_t)t$ from eq. (2.19). When time is measured in small increments, as is true in the millisecond region, the series can be terminated after the third term of eq. (2.20). Thus, in the millisecond region, the time course curve is described by the following equation.

$$(P) \simeq \tfrac{1}{2}(k_1 k_2 E_t A_t)t^2 \tag{2.21}$$

The time-course curve in this region is described by one branch of a parabola whose vertex is at the origin. This is shown in the Figure 2.1.

As time increases into the second and minute range, the exponential term in eq. (2.20) becomes negligible compared to -1, and the equation becomes

$$(P) \simeq \left(\frac{k_1 k_2 E_t A_t}{k_{-1} + k_2 + k_1 A_t}\right)t - \frac{k_1 k_2 E_t A_t}{(k_{-1} + k_2 + k_1 A_t)^2} \tag{2.22}$$

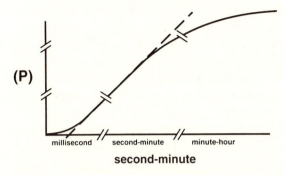

second-minute

Fig. 2.1. Time-course curve of an enzyme-catalyzed reaction. _____ The observed curve. _____ The linear relationship predicted from eq. (2.22).

Equation (2.22) defines product concentration as a linear function of time in this portion of the time course curve. This is shown in Figure 2.1 as the dashed line. The reader should recognize that the slope of this straight line is eq. (1.18), the Michaelis equation in the coefficient form. The difficulty is that eq. (2.22) predicts that the rate of the reaction will remain constant forever! This certainly does not happen. The problem lies in the fact that eq. (2.13) is the *approximate* differential equation for the EA complex. It was based on the assumption that the substrate concentration, A_t, would remain constant and that the product concentration would remain equal to zero. Neither of these assumptions apply to an actual experiment. In an actual experiment, the concentration of the substrate deceases and the concentration of the product increases until equilibrium is reached. At that point the net rate of the reaction is zero. For these latter reasons the time-course curve does not remain linear, but rather bends downward with increasing time.

2.4 A precise definition of steady-state velocity

The actual time course curve is represented in Figure 2.1 as the solid line. The slope of the curve decreases in the minute to hour region. Since the initial portion of the time course curve is concave upward and the later portion of the curve is concave downward, the curve obviously passes through an inflection point. The rate of the reaction is maximal at the inflection point of the time course curve. The rate of the reaction at that point is expressed precisely in eq. (1.18). This is the reaction rate predicted by the Michaelis equation which is based on the steady state assumption. Hence, the estimate of the rate of the reaction based on the steady state approximation is precise for only an instant, at the inflection point of the time course curve. However, it should be recalled that the second derivative of any curve is equal to zero in only two instances. The second derivative is equal to zero at any inflection point that the curve might pass through, and the second derivative is also equal to zero through any linear section of the curve. The significance of this fact is that the steady state approximation, as expressed in the Michaelis equation, is an acceptable estimate of the rate of the enzymic reaction any time the second derivative of the time course curve is indistinguishable from zero. Walter[2,3] has developed a numerical method by which the slope of the time course curve is determined at a point where the second derivative is indistinguishable from zero within the experimental error. This represents the most satisfactory method of estimating the steady state velocity of the enzyme-catalyzed reaction.

When an enzyme assay is conducted and the time course of the reaction is plotted in the second to minute range, the plot will usually exhibit the product concentration to be a linear function of time which passes through the origin. The reason for this is that the rate observed is steady state rate. The amount of product formed during the pre-steady state rate is too small to show a deviation from linearity. It is of utmost importance to recognize that the rate observed during this period of apparent linearity is a measure of the steady state rate of the reaction *only* if the conditions mandated by eqs. (1.12) and (1.13) are satisfied.

2.5 Problems for chapter 2

2.1   Show how estimates of the rate constants k_1, k_{-1} and k_2 might be obtained for the enzyme model under consideration in this chapter by using a combination of steady state and pre-steady state kinetics.

2.2   Calculate the percentage of total substrate that would be converted to product per minute if the A_t/E_t ratio were 75,000 for an enzyme with a turnover number of 100,000/minute if the substrate concentration were $0.05 \times K_m$.

2.3   Calculate the percentage of total substrate that would be converted to product per minute for the situation described in problem 2.2 if the A_t/E_t ratio were 25,000.

References

1. Reiner, J. M. (1959). *Behavior of enzyme systems.*, pp. 54–63, Minneapolis, Burgess Publ. Co.

2. Walter, C. (1966). The practicality of the use of the steady-state approximation and the inflection point method in enzyme kinetics. *J. Theor. Biol.* **15**: 1–33.
3. Walter, C. (1970). The practicality of using a power series approximation to obtain pre-steady state information. *Enzymologia* **38**: 133–39.

3

Enzyme inhibition

The significance and implications of the assumptions stated in eqs. (1.12), (1.13) and (1.14) have been treated in chapters 1 and 2. It would be reasonable to consider the fourth assumption at this point. However, the product of an enzyme-catalyzed reaction functions as an inhibitor, and since there is some variation in the terminology applied to enzyme inhibition, this chapter will be devoted to a general treatment of inhibition and will establish the terminology that will be employed in this text. Enzyme inhibition is a reversible process, and therefore does not include enzyme inactivation, for inactivation is essentially an irreversible process.

A general model of inhibition will be developed in this chapter. However, it must be borne in mind that terms such as competitive and uncompetitive inhibition refer to a particular kinetic behavior of the enzyme and not to any specific mechanism. A given mechanism may be *consistent* with a particular kinetic behavior, but there may be other mechanisms which are also consistent with the kinetic behavior.

3.1 A general model of enzyme inhibition

The following will serve as a model of enzyme inhibition.

The catalytic cycle in Figure 3.1 consists of the reactions whose rate constants are k_1, k_{-1} and k_2 and involves two enzyme species, namely E and EA. The model also includes two dead end complexes. The dead end complexes are EI and EAI. In order to derive a rate equation for the model in Figure 3.1, the following assumptions will be made.

$$E_t = (E) + (EA) + (EI) + (EAI) \qquad (3.1)$$

$$A_t \gg E_t \qquad (3.2)$$

30

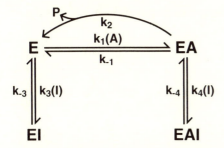

Fig. 3.1. A general model of enzyme inhibition. The free enzyme is E, EA is a binary enzyme-substrate complex, EI is the enzyme-inhibitor complex, and EAI is the ternary enzyme-substrate-inhibitor complex.

$$I_t \gg E_t \tag{3.3}$$

$$\frac{d(E)}{dt} = \frac{d(EA)}{dt} = \frac{d(EI)}{dt} = \frac{d(EAI)}{dt} = 0 \tag{3.4}$$

$$(P) = 0 \tag{3.5}$$

In addition to the foregoing assumptions, it will be assumed that inhibition is complete. This does not imply that inhibition is total so that there is no reaction rather, it implies that if the free enzyme combines with the inhibitor to form the EI complex, the EI complex cannot combine with the substrate to form the EAI complex. Furthermore, it will be assumed that the EAI complex cannot decompose to form the product. These latter assumptions are made at this time for convenience. They are not mandatory, in fact, in chapter 11 a similar model will be considered in which these assumptions will not be imposed. Equation (3.3) implies that the total inhibitor concentration is essentially equal to the free inhibitor concentration because only a negligible portion of the inhibitor is assumed to be bound to the enzyme. The rate of the enzyme-catalyzed reaction is $v = k_2(EA)$.

The steady state equations for the model are,

$$\frac{d(E)}{dt} = -k_1(E)(A) + (k_{-1} + k_2)(EA) - k_3(E)(I) + k_{-3}(EI) = 0 \tag{3.6}$$

$$\frac{d(EA)}{dt} = k_1(E)(A) + (k_{-1} + k_2)(EA) - k_4(EA)(I) + k_{-4}(EAI) = 0 \tag{3.7}$$

$$\frac{d(EI)}{dt} = k_3(E)(I) - k_3(EI) = 0 \tag{3.8}$$

$$\frac{d(EAI)}{dt} = k_4(EA)(I) - k_4(EAI) = 0 \tag{3.9}$$

The equation for the concentration of the EI complex is obtained easily,

$$\bar{K}_3 = \frac{k_{-3}}{k_3} = \frac{(E)(I)}{(EI)}, \quad (EI) = \frac{(E)(I)}{\bar{K}_3} \tag{3.10}$$

In like manner, the equation for the concentration of the EAI complex is,

$$\bar{K}_4 = \frac{k_{-4}}{k_4} = \frac{(EA)(I)}{(EAI)}, \quad (EAI) = \frac{(EA)(I)}{\bar{K}_4} \tag{3.11}$$

Equation (3.8) can be combined with eq. (3.6) to give the expression for the free enzyme complex.

$$(E) = \frac{(k_1 + k_2)(EA)}{k_1(A)} = \frac{K_m(EA)}{(A)} \tag{3.12}$$

Equations (3.10), (3.11) and (3.12) can be substituted into eq. (3.1) to give,

$$E_t = (EA)\left[1 + \frac{K_m}{(A)} + \frac{K_m(I)}{\bar{K}_3(A)} + \frac{(I)}{\bar{K}_4}\right] \tag{3.13}$$

Equation (3.13) can be solved for (EA), and multiplication by k_2 gives the rate equation for the enzymic reaction under consideration.

$$v = \frac{V_{max}}{K_m\left[1 + \frac{(I)}{\bar{K}_3}\right]\frac{1}{(A)} + 1 + \frac{(I)}{\bar{K}_4}} = \frac{V_{max}(A)}{K_m\left[1 + \frac{(I)}{\bar{K}_3}\right] + (A)\left[1 + \frac{(I)}{\bar{K}_4}\right]} \tag{3.14}$$

It was stated earlier that the Lineweaver-Burk plot is not the most satisfactory method for obtaining quantitative estimates of steady state kinetic parameters but, when such estimates have been obtained by a more suitable method, there is nothing wrong with employing Lineweaver-Burk plots in the further analysis of the data. The Lineweaver-Burk equation for the enzymic reaction is

$$\frac{1}{v} = \frac{K_m}{V_{max}}\left[1 + \frac{(I)}{\bar{K}_3}\right]\frac{1}{(A)} + \frac{1}{V_{max}}\left[1 + \frac{(I)}{\bar{K}_4}\right] \tag{3.15}$$

Substrate-saturation experiments should be conducted at each of several concentrations of the inhibitor. One of the inhibitor concentrations may be equal to zero. The experiments at each inhibitor concentration will provide an apparent Michaelis constant and an apparent maximal velocity.

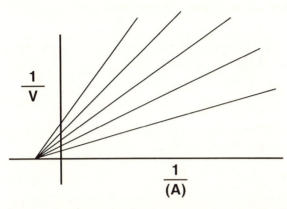

Fig. 3.2. Lineweaver-Burk plots of mixed type inhibition when the free enzyme and the EA complex have equal affinity for the inhibitor. Each line represents a substrate-saturation curve at a different inhibitor concentration.

These apparent constants will be

$$K_m^{app} = K_m \left[\frac{1 + \dfrac{(I)}{\bar{\bar{K}}_3}}{1 + \dfrac{(I)}{\bar{\bar{K}}_4}} \right] \tag{3.16}$$

$$V_{max}^{app} = \frac{V_{max}}{1 + \dfrac{(I)}{\bar{\bar{K}}_4}} \tag{3.17}$$

The plots shown in Figure 3.2 show the results of such a series of experiments for the reaction model under consideration. The inhibitor in the foregoing figure affects both the slopes and the intercepts of the Lineweaver-Burk plots. Any time an inhibitor affects both the slope and the intercept of the Lineweaver-Burk plot, that inhibitor will be classified as a mixed type inhibitor throughout this book. *Mixed type inhibition* is often referred to as noncompetitive inhibition, but some authors[1] employ a more restricted definition for noncompetitive inhibition. It is important to recognize that the term mixed type inhibition does *not* imply that this type of inhibition is a mixture of different types of inhibition. Mixed type inhibition is a distinct type of inhibition which is characterized by the affect of the inhibitor on the slope and intercept of the Lineweaver-Burk plot.

3.2 Quantitative estimates of steady state parameters in enzyme inhibition

An inspection of eq. (3.15) reveals that the slopes of the Lineweaver-Burk plots are linear functions of inhibitor concentration.

$$\text{slopes} = \frac{K_m}{V_{max} \bar{K}_3}(I) + \frac{K_m}{V_{max}} \qquad (3.18)$$

The intercepts of the primary Lineweaver-Burk plots are also linear functions of inhibitor concentration.

$$\text{intercepts} = \frac{1}{V_{max} \bar{K}_4}(I) + \frac{1}{V_{max}} \qquad (3.19)$$

Secondary plots of slopes and intercepts versus inhibitor concentration appear as shown in Figure 3.3. An analysis of the secondary plots shown in Figure 3.3 permits quantitative estimates of $\bar{K}_3$ and $\bar{K}_4$ as well as estimates of the true Michaelis constant and true maximal velocity. The primary data from the substrate-saturation experiments are used to obtain estimates of the apparent Michaelis constants and apparent maximal velocities as well as the slopes and intercepts of the primary lines in the primary plot, Figure 3.2. The true Michaelis constant and true maximal velocity are designated as K_m and V_{max}.

It is apparent from eq. (3.15) and Figure 3.2 that the family of lines which constitute the Lineweaver-Burk plots from a series of experiments at different inhibitor concentrations all intersect at some point. The coordinates of the point of intersection can be calculated by setting the Lineweaver-Burk equations for any two of the lines equal. For convenience, we may choose the Lineweaver-Burk equation in the absence of an inhibitor and that of any other line.

$$\frac{K_m}{V_{max}}\frac{1}{(A)} + \frac{1}{V_{max}} = \frac{K_m}{V_{max}}\frac{1}{(A)} + \frac{K_m(I)}{V_{max}\bar{K}_3}\frac{1}{(A)} + \frac{1}{V_{max}} + \frac{(I)}{V_{max}\bar{K}_4}$$

$$\frac{1}{(A)} = -\frac{\bar{K}_3}{K_m\bar{K}_4} \qquad (3.20)$$

This coordinate of the point of intersection on the $1/(A)$ axis can be substituted into the Lineweaver-Burk equation in the absence of an inhibitor.

$$\frac{1}{v} = \frac{K_m}{V_{max}}\left(-\frac{\bar{K}_3}{K_m\bar{K}_4}\right) + \frac{1}{V_{max}} = \frac{1}{V_{max}}\left(1 - \frac{\bar{K}_3}{\bar{K}_4}\right) \qquad (3.21)$$

This is the coordinate on the $1/v$ axis of the intersection of the family of lines on the primary plot.

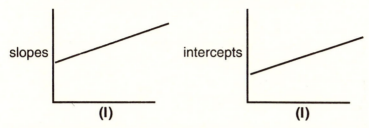

Fig. 3.3. Secondary plots of slopes and intercepts of the primary Lineweaver-Burk plots as functions of inhibitor concentration.

3.3 Competitive inhibition: A limiting case of inhibition

If $\bar{K}_3 = \bar{K}_4$, eqs. (3.20) and (3.21) show that the Lineweaver-Burk lines will intersect on the $1/(A)$ axis at a coordinate of $-1/K_m$. This is the situation portrayed in Figure 3.2. However, if the EA complex were to have a lower affinity for the inhibitor than the free enzyme, $\bar{K}_4$ would be larger than $\bar{K}_3$, and the point of intersection of the lines would be above the $1/(A)$ axis and closer to the $1/v$ axis. This is portrayed in Figure 3.4. The inhibitor in Figure 3.4 affects both the slopes and the intercepts of the Lineweaver-Burk plots, and so this is classified as mixed type inhibition even though the slopes and intercepts are not affected to the same extent by the presence of the inhibitor.

Suppose that the EA complex not only had less affinity for the inhibitor than the free enzyme, but suppose the EA complex had *no* affinity for the inhibitor. In that case, K_4, the dissociation constant of the EAI complex

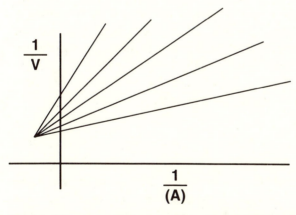

Fig. 3.4. Lineweaver-Burk plots of mixed type inhibition when the EA complex has less affinity for the inhibitor than does the free enzyme.

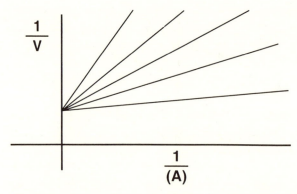

Fig. 3.5 Lineweaver-Burk plots when the EA complex has no affinity for the inhibitor, and therefore, the inhibitor is competitive with the substrate.

would be equal to infinity. The rate equation for the reaction would be,

$$v = \frac{V_{max}(A)}{K_m\left[1 + \dfrac{(I)}{\bar{\bar{K}}_3}\right] + (A)}$$

(3.22)

The Lineweaver-Burk equation is,

$$v = \frac{K_m}{V_{max}}\left[1 + \frac{(I)}{\bar{\bar{K}}_3}\right]\frac{1}{(A)} + \frac{1}{V_{max}}$$

(3.23)

This type of inhibition is called competitive. It is called competitive because the inhibitor reacts with only the same species of the enzyme as does the variable substrate. It is a limiting case of the general model shown in Figure 3.1. Figure 3.5 shows the type of Lineweaver-Burk plots that one observes with competitive inhibition. The lines in Figure 3.5 intersect on the $1/v$ axis at a coordinate of $1/V_m$. The inhibitor does not affect the intercept of the Lineweaver-Burk plots. The equation for the apparent Michaelis constant in the case of competitive inhibition is,

$$K_m^{app} = K_m\left[1 + \frac{(I)}{\bar{\bar{K}}_3}\right]$$

(3.24)

Thus, the concentration of substrate required to saturate the enzyme will be greater in the presence of a competitive inhibitor, but once the enzyme is saturated with the substrate, the inhibitor no longer exerts an affect on the kinetic behavior of the reaction.

3.4 Uncompetitive inhibition: A different limiting case

The situation where the affinity of the free enzyme for the inhibitor is less than that of the EA complex must be considered. In this case, $\bar{K}_3$ is greater than $\bar{K}_4$, and eqs. (3.20) and (3.21) show that the point of intersection of the Lineweaver-Burk lines will shift to more negative values on both axes. This will still be referred to as mixed type inhibition. However, if the free enzyme had no affinity for the inhibitor, $\bar{K}_3$ would be equal to infinity, and the point of intersection of the Lineweaver-Burk lines would be – infinity. Lines which intersect at infinity are parallel lines. This is shown in Figure. 3.6 Laidler and Bunting[2] referred to this type of inhibition as anticompetitive and their terminology is more appropriate, but most texts refer to this as uncompetitive and this less desirable term will be perpetuated in this text. Competitive and uncompetitive inhibition are the two extreme limits of the general model of inhibition shown in Figure 3.1. The rate equation for uncompetitive inhibition is given in eq. (3.25).

$$v = \frac{V_{max}(A)}{K_m + (A)\left[1 + \dfrac{(I)}{\bar{K}_4}\right]} \tag{3.25}$$

The Lineweaver-Burk equation for uncompetitive inhibition is,

$$\frac{1}{v} = \frac{K_m}{V_{max}}\frac{1}{(A)} + \frac{1}{V_{max}}\left[1 + \frac{(I)}{\bar{K}_4}\right] \tag{3.26}$$

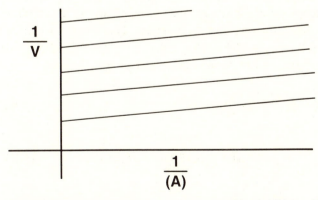

Fig. 3.6. Lineweaver-Burk plots of uncompetitive inhibition. The free enzyme has no affinity for the inhibitor.

An uncompetitive inhibitor affects the intercept, but not the slope, of a Lineweaver-Burk line. The apparent Michaelis constant in this type of inhibition is,

$$K_m^{app} = \frac{K_m}{1 + \dfrac{(I)}{\bar{K}_4}}$$

(3.27)

The apparent maximal velocity is,

$$V_{max}^{app} = \frac{V_{max}}{1 + \dfrac{(I)}{\bar{K}_4}}$$

(3.28)

An uncompetitive inhibitor decreases the Michaelis constant and the maximal velocity by the same amount and for this reason the slope of the Lineweaver-Burk plot is not affected.

3.5 Substrate inhibition

One additional type of inhibition should be discussed before leaving the general topic of enzyme inhibition. There are numerous types of substrate inhibition, only one of which will be considered at this point. Other types of substrate inhibition will be discussed in subsequent sections of this book. The type of substrate inhibition which is considered here can be visualized as an enzyme which has multiple sites of attachment for the substrate. If the concentration of the substrate is sufficiently high, more than one substrate molecule may occupy the active site at one time which would result in the formation of an unproductive ternary complex. This situation can be visualized in the following model.

The following equations can be written for each enzyme species,

$$\frac{d(E)}{dt} = -k_1(E)(A) + (k_{-1} + k_2)(EA) = 0$$

(3.29)

$$\frac{d(EA)}{dt} = k_1(E)(A) - (k_{-1} + k_2)(EA) - k_i(EA)(A) + k_{-i}(AEA) = 0$$

(3.30)

$$\frac{d(AEA)}{dt} = k_i(EA) - k_{-i}(AEA) = 0$$

(3.31)

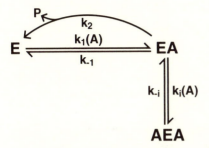

Fig. 3.7. Model of substrate inhibition when two molecules of substrate bind to the active site to form an unproductive AEA complex.

Equations (3.29) and (3.31) can be rearranged.

$$(E) = \frac{K_m(EA)}{(A)} \tag{3.32}$$

$$(AEA) = \frac{(EA)(A)}{\bar{K}_i} \tag{3.33}$$

In the foregoing equations $K_i = k_{-i}/k_i$. The enzyme conservation equation for the model in Figure 3.7 is,

$$E_t = (E) + (EA) + (AEA) = (EA)\left[1 + \frac{K_m}{(A)} + \frac{(A)}{\bar{K}_i} \right] \tag{3.34}$$

Since the rate of the reaction is $v = k_2(EA)$, the equation for the velocity of the reaction is obtained by solving eq. (3.34) for the concentration of the EA complex and multiplying the result by k_2.

$$v = \frac{V_{max}}{K_m + (A)\left[1 + \frac{(A)}{\bar{K}_i} \right]} = \frac{V_{max}(A)}{K_m + (A) + \frac{(A)^2}{\bar{K}_i}} \tag{3.35}$$

The previous rate equations, which were derived in this chapter and in chapter 1, were rational polynomials of order 1:1, that is, they described a rectangular hyperbola. While eq. (3.35) is a rational polynomial it is a 1:2 function. The numerator contains a substrate concentration to the first power while the denominator contains a substrate concentration to the second power. The rate of the reaction in eq. (3.35) is not a hyperbolic function of the substrate concentration. The substrate-saturation curve will appear as in Figure 3.8. In many actual experiments, the substrate inhibition may not be as acute as portrayed in Figure 3.8, and it may not be

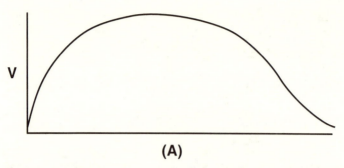

(A)

Fig. 3.8. Substrate-saturation curve for the model of substrate inhibition portrayed in Figure 3.7.

possible to extend the experiment to include as broad a substrate concentration range. Nevertheless, it should be noted that eq. (3.35) predicts that the rate of the reaction will approach a value of zero in an asymptotic manner as (A) approaches infinity.

The Lineweaver-Burk equation for the model shown in Figure 3.7 is given in eq. (3.36).

$$\frac{1}{v} = \frac{K_m}{V_{max}}\frac{1}{(A)} + \frac{1}{V_{max}}\left[1 + \frac{(A)}{\bar{\bar{K}}_i}\right] \tag{3.36}$$

The Lineweaver-Burk plot would have the appearance of Figure 3.9. The distinctive feature about the Lineweaver-Burk plot, in the case of substrate inhibition, is that the curve bends upward sharply as it approaches the $1/v$ axis. The curve approaches a straight line in an asymptotic manner as the

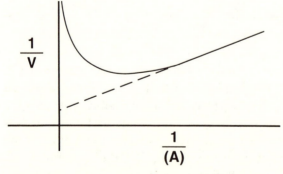

$$\frac{1}{(A)}$$

(A)

Fig. 3.9. Lineweaver-Burk plot of the substrate-saturation curve for the model in Figure 3.7. Extrapolation of the linear portion of the curve at high values of $1/(A)$ _____.

values of $1/(A)$ become large. It is customary to extrapolate this line to the $1/v$ axis and interpret this intercept as $1/V_{max}$ and to interpret the slope as K_m/V_{max}. However, eq. (3.35) can be rearranged as,

$$\frac{(A)}{v} = \frac{1}{V_{max}\bar{K}_i}(A)^2 + \frac{1}{V_{max}}(A) + \frac{K_m}{V_{max}} \tag{3.37}$$

A Lineweaver-Burk plot is a good *qualitative* indicator of substrate inhibition, but if there is good reason to believe that Figure 3.7 is an appropriate model for the enzymic reaction, then eq. (3.37) provides a better means of obtaining *quantitative* estimates of the kinetic parameters because eq. (3.37) is a simple second order polynomial when $(A)/v$ is plotted against (A). The appendix to this chapter contains a brief description of the regression analysis of a second order polynomial.

As mentioned previously, Figure 3.7 is one model which will provide for substrate inhibition. Other models of substrate inhibition will be discussed in subsequent chapters.

3.6 Problems for chapter 3

3.1 The following data were obtained in a series of substrate-saturation experiments conducted with various concentrations of an inhibitor. Plot these and obtain estimates of the apparent Michaelis constants and maximum velocities for each concentration of the inhibitor, and obtain estimates of the true Michaelis constant, the true maximum velocity and estimates of the inhibition constants from the secondary plots. In the following data, the velocities are expressed as μmoles per minute.

(A) mM	(I) mM 0.000	(I) mM 1.393	(I) mM 2.790	(I) mM 4.180	(I) mM 5.570
	v	v	v	v	v
0.05	2.62	1.77	1.40	1.29	
0.10	4.54	3.00	2.32	2.10	
0.15	5.72	3.84	3.00	2.66	2.50
0.20	6.67	4.77	3.50	3.11	2.79
0.25	7.43	5.00	4.06	3.55	3.19
0.30	7.90	5.42	4.48	3.74	3.53
0.40	9.04	6.07	4.80	4.17	3.83

3.2   Rearrange eq. (3.14) such that (I) can be plotted as a function of $(v_0 - v_i)/v_i$, where v_0 is the velocity of the enzymic reaction in the absence of the inhibitor and v_i is the velocity in the presence of the inhibitor. Note that in this relationship, when the enzyme is subject to 50 percent inhibition, the value of $(v_0 - v_i)/v_i$ is equal to unity.

3.3   Observe the appearance of the relationship derived in problem 3.2 when each of the following conditions apply;

(a) $\bar{K}_3 = \bar{K}_4$, (b) $K_m/(A) = 0$, (c) $\bar{K}_4 = \infty$, (d) $\bar{K}_4 = \infty$ and (A) $= K_m$, (e) $\bar{K}_3 = \infty$, and (f) $\bar{K}_3 = \infty$ and $K_m/(A) = 0$.

Appendix: A brief discussion of polynomial regression

Polynomial regression analysis proceeds in a manner similar to linear regression[3,4]. Second order polynomial regression serves as an example. The equation for a second order polynomial is

$$Y = b_0 + b_1 X + b_2 X^2 \tag{3.A.1}$$

The aim of regression is to obtain estimates of the parameters b_0, b_1 and b_2 which describe a regression line which fits the experiment data best. The assumptions listed in the appendix of chapter 1 will apply in this treatment of regression analysis of a second order polynomial. For any value of the independent variable, the observed value of the dependent variable will deviate from the regression line by a residual equal to $Y - \hat{Y}$, where $\hat{Y}$ is the expected value from the regression equation. The procedure is to obtain estimates of the parameters which will minimize the sum of the squared residuals.

$$\sum r^2 = \sum (Y - b_0 - b_1 X - b_2 X^2)^2 \tag{3.A.2}$$

The expansion of the right-hand side of eq. (3.A.2) followed by differentiation with respect to each of the parameters gives,

$$\frac{d\sum r^2}{db_0} = -2\sum Y + 2nb_0 + 2b_1 \sum X + 2b_2 \sum X^2 \tag{3.A.3}$$

$$\frac{d\sum r^2}{db_1} = -2\sum XY + 2b_0 \sum X + 2b_1 \sum X^2 + 2b_2 \sum X^3 \tag{3.A.4}$$

$$\frac{d\sum r^2}{db_2} = -2\sum X^2 Y + 2b_0 \sum X^2 + 2b_1 \sum X^3 + 2b_2 \sum X^4 \tag{3.A.5}$$

Minimization of the sum of the squared residuals with respect to the parameters is achieved by setting each of the foregoing three equations equal to zero and solving the system of equations for the parameters. The resultant equations are,

$$\sum Y = nb_0 + \sum X b_1 + \sum X^2 b_2 \tag{3.A.6}$$

$$\sum XY = \sum X b_0 + \sum X^2 b_1 + \sum X^3 b_2 \tag{3.A.7}$$

$$\sum X^2 Y = \sum X^2 b_0 + \sum X^3 b_1 + \sum X^4 b_2 \tag{3.A.8}$$

The following augmented matrix can be constructed for the system of equations.

$$\begin{vmatrix} n & \sum X & \sum X^2 & \sum Y & 1 & 0 & 0 \\ \sum X & \sum X^2 & \sum X^3 & \sum XY & 0 & 1 & 0 \\ \sum X^2 & \sum X^3 & \sum X^4 & \sum X^2 Y & 0 & 0 & 1 \end{vmatrix}$$

The terms which must be obtained from the experimental data are

$$n, \sum x, \sum X^2, \sum X^3, \sum X^4, \sum Y, \sum Y^2, \sum XY, \sum X^2 Y.$$

Matrix inversion can be conducted by the Gauss-Jordan elimination as outlined in the appendix of chapter 1[5].

The resultant augmented matrix will be

$$\begin{vmatrix} 1 & 0 & 0 & b_0 & c_{11} & c_{12} & c_{13} \\ 0 & 1 & 0 & b_1 & c_{21} & c_{22} & c_{23} \\ 0 & 0 & 1 & b_2 & c_{31} & c_{32} & c_{33} \end{vmatrix}$$

where c_{ij} are the elements of the inverted matrix.

The values of the parameters for the regression equation of a second order polynomial are the elements of the fourth column, or in more general terms, the elements of the $m + 2$ column where m is the order of the polynomial. The variance for a second order polynomial is given by

$$s^2 = \frac{\sum Y^2 - b_0 \sum Y - b_1 \sum XY - b_2 \sum X^2 Y}{n - 3} \tag{3.A.9}$$

The standard error of the estimates of the parameters is given by the general equation

$$[\text{s.e.}_{b_{i-1}} + \sqrt{s^2 \times c_{ii}}]_{i=1,2,\ldots,m} \tag{3.A.10}$$

Polynomial regression is applicable to a plot of $(A)/v$ vs. (A) when substrate inhibition gives rise to a rate equation which is a $1:2$ function of substrate concentration. In later chapters, enzyme models which result in rate equations which are $2:2$ functions will be considered. This present type of analysis is not appropriate in those cases.

References

1. Segel, I. H. (1975). *Enzyme Kinetics*, pp. 100–226, New York, John Wiley & Sons.
2. Laidler, K. J. and Bunting, P. S. (1973). *The Chemical Kinetics of Enzyme Action.*, 2d. Ed., pp. 68–113, Oxford, Clarendon Press.
3. Steele, R. G. D. and Tories, J. H. (1960). *Principles and Procedures of Statistics*, pp. 338–40, New York, McGraw-Hill.
4. Carnahan, B., Luther, H. A. and Wilkes, J. O. (1969). *Applied Numerical Methods*, pp. 573, New York, John Wiley & Sons.
5. Carnahan, B., Luther, H. A. and Wilkes, J. O. (1969). *Applied Numerical Methods*, pp. 269–73, New York, John Wiley & Sons.

4
Reversible enzyme-catalyzed reactions

The derivation of eq. (1.18), a rate equation for an enzyme-catalyzed reaction, was possible because a number of assumptions were stipulated. These assumptions were expressed in eqs. (1.12) through (1.15). The first three of these assumptions are mandatory, and they have been discussed in some detail in chapters 1 and 2. The fourth assumption, namely (P) = 0, was imposed as a matter of convenience, and it is not an absolute requirement. The thrust of the present chapter will be to investigate the kinetic behavior of an enzyme-catalyzed reaction when the restriction (P) has been removed.

4.1 Derivation of a rate equation by matrix inversion

A model for such a reaction is shown in Figure 4.1. There is a logical problem with the sequence shown in Figure 4.1. As the reaction proceeds from left to right, the enzyme combines with the substrate to form an enzyme-substrate complex which can either dissociate to enzyme plus substrate or be converted to enzyme plus product. However, when the reaction proceeds from right to left, the model shows that the enzyme combines with the product to form an enzyme-substrate complex instantaneously. Figure 4.2 presents a sequence which seems more logical. Whether or not the reaction sequences shown in Figure 4.1 and Figure 4.2 are distinct on the basis of steady state kinetics, the latter sequence is more acceptable logically. Yet, the question still arises, "Might the reaction sequence involve even more intermediate binary complexes?" If so, would this affect the steady state behavior of the enzyme? One way to approach this problem is to derive the steady state rate equation for the reaction sequence in Figure 4.2, and then set the product concentration equal to zero and see how that equation compares to eq. (1.21), which was obtained for the reaction sequence in Figure 1.1.

$$E + A \underset{k_{-1}}{\overset{k_1}{\rightleftharpoons}} EA \underset{k_{-2}}{\overset{k_2}{\rightleftharpoons}} E + P$$

Fig. 4.1. Model of a reversible single substrate, single product enzyme-catalyzed reaction with a single binary EA complex.

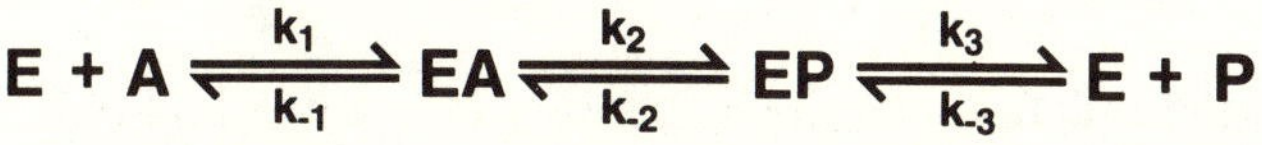

$$E + A \underset{k_{-1}}{\overset{k_1}{\rightleftharpoons}} EA \underset{k_{-2}}{\overset{k_2}{\rightleftharpoons}} EP \underset{k_{-3}}{\overset{k_3}{\rightleftharpoons}} E + P$$

Fig. 4.2. Model of a reversible single substrate, single product enzyme-catalyzed reaction which contains an EA and an EP complex.

The assumption will be made that the amount of substrate and product bound to the enzyme is negligible so that $A_t = (A)$ and $P_t = (P)$. The enzyme conservation equation is,

$$E_t = (E) + (EA) + (EP) \tag{4.1}$$

The steady state equations for the three enzyme complexes are,

$$\frac{d(E)}{dt} = -[k_1(A) + k_{-3}(P)](E) + k_{-1}(EA) + k_3(EP) = 0 \tag{4.2}$$

$$\frac{d(EA)}{dt} = k_1(A)(E) - (k_{-1} + k_2)(EA) + k_{-2}(EP) = 0 \tag{4.3}$$

$$\frac{d(EP)}{dt} = k_{-3}(P)(E) + k_2(EA) - (k_{-2} + k_3)(EP) = 0 \tag{4.4}$$

Equations (4.2) through (4.4) can be written in matrix form as,

$$\begin{vmatrix} -[k_1(A) + k_{-3}(P)] & k_{-1} & k_3 \\ k_1(A) & -(k_{-1} + k_2) & k_{-2} \\ k_{-3}(P) & k_2 & -(k_{-2} + k_3) \end{vmatrix} \begin{vmatrix} (E) \\ (EA) \\ (EP) \end{vmatrix} = \begin{vmatrix} 0 \\ 0 \\ 0 \end{vmatrix} \tag{4.5}$$

For those readers who are not familiar with matrix algebra, the book by Magar[1] contains a concise discussion of the basic rules of matrix algebra. Equation (4.5) can be expressed simply as,

$$AB = C \tag{4.6}$$

The element in the first row of C is the sum of the first element of the first row of A times the element of the first row of B, plus the second element of the first row of A times the element of the second row of B, plus the third element of the first row of A times the element of the third row of B. Likewise, the element of the second row of C is the sum of the first element of the second row of A times the element of the first row of B plus the second element of the second row of A times the element of the second row of B, plus the third element of the second row of A times the element of the third row of B. The foregoing procedure is continued until all the equations are obtained.

Unfortunately, the equations which comprise eq. (4.5) are not all independent, but this can be resolved by replacing any one of the equations with eq. (4.1), the enzyme conservation equation, and that will provide an adequate number of independent equations. Stated otherwise, for an enzyme sequence containing n enzyme species, there are n steady state equations plus the enzyme conservation equation to give $n+1$ equations which will provide n independent equations which can be solved for the n enzyme species. The procedure which will be followed in this text is to replace the steady state equation for the enzyme species in question with the enzyme conservation equation. For example, to derive the expression for the concentration of the free enzyme, eq. (4.2) would be replaced by eq. (4.1). The following system of equations could be solved to obtain an expression for (E).

$$\begin{vmatrix} 1 & 1 & 1 \\ k_1(A) & -(k_{-1}+k_2) & k_{-2} \\ (k_{-3}(P) & k_2 & -(k_{-2}+k_3) \end{vmatrix} \begin{vmatrix} (E) \\ (EA) \\ (EP) \end{vmatrix} = \begin{vmatrix} E_t \\ 0 \\ 0 \end{vmatrix} \tag{4.7}$$

The expression for the concentration of the free enzyme is a quotient, the numerator of which is the 3×3 matrix of eq. (4.7) in which the first column is replaced by the column vector on the right-hand side of eq. (4.7), and the denominator is the 3×3 matrix of eq. (4.7). Thus, the concentration of the free enzyme would be given by the following quotient.

$$(E) = \frac{\begin{vmatrix} E_t & 1 & 1 \\ 0 & -(k_{-1}+k_2) & k_{-2} \\ 0 & k_2 & -(k_{-2}+k_3) \end{vmatrix}}{\begin{vmatrix} 1 & 1 & 1 \\ k_1(A) & -(k_{-1}+k_2) & k_{-2} \\ k_{-3}(P) & k_2 & -(k_{-2}+k_3) \end{vmatrix}} \tag{4.8}$$

Solving eq. (4.8) for (E) requires inversion of both the numerator and denominator matrices. Stated in an analogous manner, it is necessary to obtain the symbolic determinants of these two matrices. There are a number of efficient methods for extracting numerical determinants, one of which has been outlined in the appendix of Chapter 1. However, the procedures for obtaining a numerical determinant are not convenient for obtaining a symbolic determinant. Cramer's rule is probably the most feasible method of obtaining a symbolic determinant, and Cramer's rule is not a particularly efficient method of matrix inversion. Only a square matrix has a determinant, and the determinant is the sum of all of the possible combinations of elements, where each combination consists of only one element from each row of the matrix and only one element from each column of the matrix. Some of the combinations, or permutations, are positive while others are negative. The actual sign of each permutation is determined by the sign associated with each element and also by location of the element in the matrix. It is obvious that the permutations and their sign must be determined in a systematic manner. Cramer's rule provides a systematic method of accomplishing this goal. The method of matrix inversion which will be described here, and used extensively in the last section of this text, is based on Cramer's rule, but it utilizes an algorithm which can be incorporated easily into a computer-based method[2]. The method consists of constructing a secondary matrix from the matrix to be inverted. The secondary matrix is called a Q matrix, and it consists of elements which identify the column numbers of the non-zero elements in the corresponding row of the primary matrix. Figure 4.3 shows the denominator matrix of eq. (4.8), which is identified as $|D|$, and its associated Q matrix.

Each permutation is represented as a vector. The vector is constructed such that each element of the vector is taken from a different row of Q, but since the elements of Q represent columns of a primary matrix, there can be no repetition of numbers in the vector. The allowable vectors which represent each permutation for the matrices in Figure 4.3 are given in the first column of Figure 4.4. The second column in Figure 4.4 is the sum of

$$|D| = \begin{vmatrix} 1 & 1 & 1 \\ k_1(A) & -(k_{-1} + k_2) & k_{-2} \\ k_{-3}(P) & k_2 & -(k_{-2} + k_3) \end{vmatrix}, \quad Q = \begin{vmatrix} 1 & 2 & 3 \\ 1 & 2 & 3 \\ 1 & 2 & 3 \end{vmatrix}$$

Fig. 4.3. Denominator matrix for the model in Fig. 4.2. The Q matrix is the matrix of non-zero elements in the denominator matrix.

$(1,2,3)$ $0 + 1$ $+ k_{-1}k_{-2} + k_{-1}k_2 + k_2k_{-2} + k_2k_3$
$(1,3,2)$ $1 + 0$ $- k_2k_2 - k_2k_{-2}$
$(2,1,3)$ $1 + 1$ $+ k_1\bar{k}_{-2}(A) + k_1k_3(A)$
$(2,3,1)$ $2 + 0$ $+ k_{-2}k_{-3}(P)$
$(3,1,2)$ $2 + 0$ $+ k_1k_2(A)$
$(3,2,1)$ $3 + 1$ $+ k_{-1}k_{-3}(P) + k_2k_{-3}(P)$

Fig. 4.4 Expansion of the denominator matrix shown in Fig. 4.3. The first column presents the possible permutations in vector form. The second column consists of the two integers the sum of which is equal to p. See text for further details of the p value. The last column contains the terms of the symbolic determinant.

two integers. The first of these is the number of deviations from sequence in the vector in column 1. Thus, in the sequence $1, 2, 3$ there are no deviations from sequence so the first integer in column two of row 1 is 0. On the other hand, the sequence $3, 2, 1$ in row six has three deviations from sequence because the 3 precedes 2 and 1, and 2 precedes 1. Therefore the first integer in column 2 of row 6 is 3. The second integer in column 2 is a number of negative terms in the permutation indicated by the vector in column 1. This is obtained by reference to the matrix in Figure 4.3. The sum of the integers in column 2 is termed p, and the sign of the permutation is given by the multiplication of the terms in the permutation by $(-1)^p$. Column three in Figure 4.4 consists of the terms contained in the permutation. It should be noted that all of the terms are positive except the term in the second row. Thus, k_2k_{-2} in the third column of row one is canceled by $-k_2k_{-2}$ in the third column of row two. The denominator determinant in eq. (4.8) is

$$|D| = \begin{matrix} k_{-1}k_{-2} + k_{-1}k_3 + k_2k_3 \\ + k_1k_{-2}(A) + k_1k_3(A) + k_{-1}k_{-3}(P) \\ + k_1k_2(A) + k_{-1}k_{-3}(P) + k_2k_{-3}(P) \end{matrix} \qquad (4.9)$$

The numerator matrix and its corresponding Q matrix are shown in Figure 4.5. Figure 4.6 shows the permutations, p values and terms associated with the numerator of eq. (4.8). The equation for the concentration of the free enzyme is

$$(E) = \frac{(k_1k_{-2} + k_{-1}k_2 + k_2k_3)\,E_t}{\begin{matrix} k_{-1}k_{-2} + k_{-1}k_3 + k_2k_3 \\ + k_1k_{-2}(A) + k_1k_3(A) + k_{-2}k_{-3}(P) \\ + k_1k_2(A) + k_{-1}k_{-3}(P) + k_2k_{-3}(P) \end{matrix}} \qquad (4.10)$$

$$|D| = \begin{vmatrix} E_t & 1 & 1 \\ 0 & -(k_{-1}+k_2) & k_{-2} \\ 0 & k_2 & -(k_{-2}+k_3) \end{vmatrix}, \quad Q = \begin{vmatrix} 1 & 2 & 3 \\ 2 & 3 & 0 \\ 2 & 3 & 0 \end{vmatrix}$$

Fig. 4.5 Numerator matrix for the free enzyme for the model in Fig. 4.2. The Q matrix is the matrix of non-zero elements in the numerator matrix.

Vector	P	Terms
(1,2,3)	0+2	$+ (k_{-1}k_{-2} + k_{-1}k_3 + k_2 k_{-2} + k_2 k_3)\,E_t$
(1,3,2)	1+0	$- k_2 k_{-2}\, E_t$

Fig. 4.6 Expansion of the numerator matrix in Fig. 4.5.

If both sides of eq. (4.10) are divided by E_t, the equation is an expression of the fraction of the total enzyme that is present as free enzyme in the steady state. That is, it is the distribution expression for the free enzyme. It should also be noted that the numerator of this distribution expression is equal to the first row of the denominator.

An expression for the concentration of the EA complex is obtained by replacing eq. (4.3) with the enzyme conservation equation. The equation in matrix form is

$$(EA) = \frac{\begin{vmatrix} -[k_1(A)+k_{-3}(P)] & 0 & k_3 \\ 1 & E_t & 1 \\ k_{-3}(P) & 0 & -(k_{-2}+k_3) \end{vmatrix}}{\begin{vmatrix} -[k_1(A)+k_{-3}(P)] & k_{-1} & k_3 \\ 1 & 1 & 1 \\ k_{-3}(P) & k_2 & -(k_{-2}+k_3) \end{vmatrix}} \tag{4.11}$$

The denominator determinant in eq. (4.11) is exactly the same as that given in eq. (4.9). The distribution expression for the EA complex is,

$$\frac{(EA)}{E_t} = \frac{k_1 k_2(A) + k_1 k_3(A) + k_{-2}k_{-3}(P)}{|D|} \tag{4.12}$$

The numerator of eq. (4.12) is equal to the second row of the denominator determinant. In like manner, the expression for the EP complex is obtained by replacing eq. (4.4) with the enzyme conservation equation.

$$(EP) = \cfrac{\begin{vmatrix} -[k_1(A) + k_{-3}(P)] & k_{-1} & 0 \\ k_1(A) & -(k_{-1} + k_2) & 0 \\ 1 & 1 & E_t \end{vmatrix}}{\begin{vmatrix} [k_1(A) + k_{-3}(P)] & k_{-1} & k_3 \\ k_1(A) & -(k_{-1} + k_2) & k_{-2} \\ 1 & 1 & 1 \end{vmatrix}} \qquad (4.13)$$

Once again, the denominator determinant is identical to eq. (4.9), and the equation for the distribution expression for the EP complex is given by eq. (4.14).

$$\frac{(EP)}{E_t} = \frac{k_1 k_2(A) + k_{-1} k_{-3}(P) + k_2 k_{-3}(P)}{|D|} \qquad (4.14)$$

The numerator of eq. (4.14) is identical to the third row of the expression for the denominator determinant, and this shows that the denominator determinant is equal to E_t in terms of rate constants and the concentration of the reactants. The rate of the reaction of the enzymic reaction shown in Figure 4.2 is

$$v = k_3(EP) - k_{-3}(E)(A) \qquad (4.15)$$

The expression for (EP) from eq. (4.14) and the expression for (E) from eq. (4.10) can be substituted into eq. (4.15).

$$v = \frac{[k_1 k_2 k_3(A) - k_{-1} k_{-2} k_{-3}(P)]E_t}{\begin{aligned}&k_{-1}k_{-2} + k_{-1}k_3 + k_2 k_3 + k_1(k_2 + k_{-2} + k_3)(A) \\ &\quad + k_{-3}(k_{-1} + k_2 + k_{-2})(P)\end{aligned}} \qquad (4.16)$$

Equation (4.16) is the complete rate equation in the coefficient form for the enzyme sequence shown in Figure 4.2.

4.2 Reformulation of the complete rate equation

If the concentration of the product is set equal to zero, eq. (4.16) becomes,

$$v = \frac{k_1 k_2 k_3 E_t(A)}{k_{-1}k_{-2} + k_{-1}k_2 + k_2 k_3 + k_1(k_2 + k_{-2} + k_3)(A)} \qquad (4.17)$$

Equation (4.17) appears distinctly different from eq. (1.18), but it must be recalled that steady state kinetic studies do not usually provide information about individual rate constants. Steady state kinetic studies provide information about the steady state parameters. Equation (4.16) could be re-written as

$$v = \frac{\text{num. } 1(A) - \text{num. } 2(P)}{\text{constant} + \text{coef. } A(A) + \text{coef. } P(P)} \tag{4.18}$$

If (P) is set equal to zero in eq. (4.18), the equation is identical to that obtained in chapter 1. This emphasizes that one does not know the composition of steady state parameters on the basis of experimental observations; rather, this is known only after derivation of the rate equation for a particular enzyme model. If (P) = 0, the maximal velocity in the forward direction and the Michaelis constant for A are expressed as follows.

$$V_f = \frac{\text{num. } 1}{\text{coef. } A} = \frac{k_2 k_3 E_t}{k_2 + k_{-2} + k_3} \tag{4.19}$$

$$K_a = \frac{\text{constant}}{\text{coef. } A} = \frac{k_{-1} k_{-2} + k_{-1} k_2 + k_2 k_3}{k_1 (k_2 + k_{-2} + k_3)} \tag{4.20}$$

In like manner, if (A) = 0, the maximal velocity in the reverse direction and the Michaelis constant for P are defined as,

$$V_r = \frac{\text{num. } 2}{\text{coef. } P} = \frac{k_{-1} k_{-2} E_t}{k_{-1} + k_2 + k_3} \tag{4.21}$$

$$K_p = \frac{\text{constant}}{\text{coef. } P} = \frac{k_{-1} k_{-2} + k_{-1} k_2 + k_2 k_3}{k_{-3} (k_{-1} + k_2 + k_{-2})} \tag{4.22}$$

The equilibrium constant is defined as,

$$K_{eq} = \frac{\text{num. } 1}{\text{num. } 2} = \frac{k_1 k_2 k_3 E_t}{k_{-1} k_{-2} k_{-3} E_t} \tag{4.23}$$

Equation (4.23) indicates what is already known, namely, that an enzyme does not affect the equilibrium of the catalyzed reaction. The task of reformulating the complete rate equation from the coefficient form into the more useful kinetic form stills remains. This is accomplished in the same manner as outlined in chapter 1. Each term in the numerator and denominator of eq. (4.18) is divided by the coefficient of the denominator term in the variable substrate. The variable substrate for the reaction in the forward direction is A. The reformulation process is illustrated in Figure 4.7. Substitution of eqs. (4.19) through (4.23) into the expression in the foregoing

$$v = \frac{\dfrac{num.\,1}{coef.\,A}(A) - \dfrac{num.\,2}{coef.\,A}\dfrac{num.\,1}{num.\,1}(P)}{\dfrac{constant}{coef.\,A} + \dfrac{coef.\,A}{coef.\,A}(A) + \dfrac{coef.\,P}{coef.\,A}\dfrac{constant}{constant}(P)}$$

Fig. 4.7. Reformulation of the rate equation for the model in Figure 4.2 from the coefficient form to the kinetic form.

figure provides the complete rate equation in the kinetic form.

$$v = \frac{V_f\left[(A) - \dfrac{(P)}{K_{eq}}\right]}{K_a + (A) + \dfrac{K_a}{K_p}(P)} = \frac{V_f\left[1 - \dfrac{(P)}{K_{eq}(A)}\right]}{1 + \dfrac{K_a}{(A)}\left[1 + \dfrac{(P)}{K_p}\right]} \tag{4.24}$$

An equation identical to eq. (4.24) would have been derived for the reaction sequence portrayed in Figure 4.1. While the rate equation in the coefficient form derived for the model in Figure 4.1 would differ from eq. (4.16), it is the equation in the kinetic form that is relevant to steady state kinetic studies. A principle which emerges from this is that if the interconversion of two enzyme species does *not* involve the interaction of the enzyme with a reactant, there is nothing gained by including both of the enzyme species in the reaction sequence for the purpose of deriving the steady state rate equation. It will suffice to include only one of the species. Thus, while Figure 4.2 presents a more logical reaction sequence, the steady state rate equations for the two models are identical in the kinetic form.

4.3 The effect of product inhibition

It is apparent from eq. (4.24) that the product of the reaction will decrease the rate of the reaction for two reasons. The second term in the numerator is negative and contains the concentration of the product. The extent to which the rate is decreased by this term is dependent on how close the reaction is to equilibrium. When the reaction is at equilibrium, the numerator of eq. (4.24) is equal to zero. The product will also inhibit the reaction because the third term in the denominator contains the concentration of product. The extent to which the reaction rate is decreased by this term is independent of classical thermodynamic considerations, but rather it is determined by kinetic factors. It is proportional to K_a/K_p. The product of the reaction can inhibit an enzyme-catalyzed reaction even when the reaction is infinitely far from equilibrium. For example, if k_{-2} were equal to zero in the reaction sequence portrayed in Figure 4.2, the reaction would be irreversible thermodynamically. Reference to eq. (4.23) shows that the equilibrium constant would be equal to infinity if k_{-2} were equal to 0. The steady state parameters would be somewhat affected; of particular significance, the Michaelis constants would be

$$K_a = \frac{k_{-3}(k_{-1} + k_2)}{k_1(k_2 + k_3)} \tag{4.25}$$

$$K_p = \frac{k_3}{k_{-3}} \tag{4.26}$$

Under this condition, the Michaelis constant for P becomes a dissociation constant, that is, it is a thermodynamic rather than a kinetic parameter. Unlike the equilibrium constant, it is not necessarily equal to infinity. If $k_{-2} = 0$, the rate of the reaction would be,

$$v = \frac{V_f E_t(A)}{K_a \left[1 + \dfrac{(P)}{K_p} \right] + (A)} \tag{4.27}$$

In the light of the discussion of enzyme inhibitors in the previous chapter, it is apparent that, in the case of the model portrayed in Figure 4.2, the product P would function as a competitive inhibitor. That should not be surprising, for it can be seen by reference to Figure 4.2 that both A and P react with the same species of the enzyme. Thus, even when the reaction is infinitely far from equilibrium, the product can serve as an inhibitor. In some cases the Michaelis or inhibition constant of the product may be so large that the ratio of product concentration to that of the constant may be insignificantly small, but that condition must be established experimentally. The assumption that product inhibition is negligible should never be made arbitrarily.

4.4 Use of the King-Altman method to derive the rate equation

The rate equation for an enzyme-catalyzed reaction can always be derived by matrix inversion. However, if, for example the enzyme model contained six enzyme species, then the derivation would require the repeated inversion of a 6×6 matrix. This is not an easy task if it is to be done manually. King and Altman[3] applied the graph theory and developed a graphical method which greatly simplifies the derivation. In order to facilitate this treatment, the enzyme model shown in Figure 4.2 can be shown in a cyclic

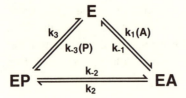

Fig. 4.8 Cyclic representation of the enzymic model in Figure 4.2.

rather than a linear sequence. Figure 4.8 can be looked upon as a graph in which each enzyme species is a vertex (node) and the vertices are connected by lines (edges). The edges have arrows to indicate direction, thus Figure 4.8 is a directed graph (digraph). The rate constants of the reactions times the concentration of any reactant that interacts with the enzyme are weighting factors associated with the appropriate edge. In graph theory, a digraph as shown in Figure 4.8 can represent a system of equations[4]. Furthermore, a spanning tree is a non-cyclic path which connects all the graph vertices to one of the vertices. The significance of the stipulation, non-cyclic, is that no vertex can be encountered more than once in each path. King and Altman showed that the equation for the distribution of any enzyme species in an enzymic reaction is equal to the sum of all of the spanning trees leading to that enzyme species divided by the sum of all of the spanning trees in the digraph. The spanning trees for (E) are shown in Figure 4.9. Figure 4.9 gives the numerator of the distribution expression for the free enzyme. The denominator is the sum of all of the spanning trees, and this is identical to eq. (4.9). It should be noted that any path that contains a sequence of rate constants that contain $k_i k_{-i}$ will certainly constitute a cyclic path and therefore will not be a valid spanning tree. These paths always cancel out during the matrix inversion operation shown earlier in this chapter. The spanning trees shown in Figure 4.10 are those for the distribution expression for (EA), and they are identical to the numerator of eq. (4.12). The spanning trees in Figure 4.11 are those for the enzyme-product complex.

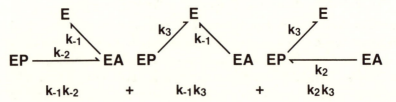

Fig. 4.9. The spanning trees which constitute the King-Altman solution for the free enzyme in Figure 4.8.

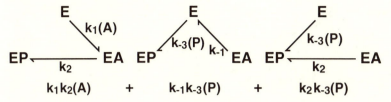

Fig. 4.10. The spanning trees which constitute the King-Altman solution for the EA complex in Figure 4.8.

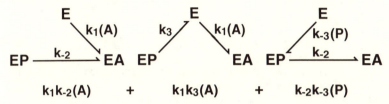

Fig. 4.11. The spanning trees which constitute the King-Altman solution for the EP complex in Figure 4.8.

The King-Altman graphical approach to derivation of the rate equation for an enzyme-catalyzed reaction is far simpler than the matrix inversion procedure discussed earlier in this chapter. This demonstrates the power of graph theory for this derivation. The difference between the King-Altman method and actual matrix inversion would be even more dramatic in the case of more complex reaction sequences. The reader who is interested in other applications of the King-Altman method is referred to an excellent series of publications by Terrell Hill[5-9] and also papers by Poland[10] and Chou[11]. However, the matrix inversion procedure is more general, and it will be employed extensively in the section on multi-enzyme systems.

4.5 Problems for chapter 4

4.1   Derive the rate equation for the following enzyme model (Fig. 4.12) and reformulate it to the kinetic form.

4.2   Derive the rate equation for the following enzyme reaction sequence (Fig. 4.13) in the coefficient form using the King-Altman method.

$$E + A \underset{k_{-1}}{\overset{k_1}{\rightleftharpoons}} EA \underset{k_{-2}}{\overset{k_2}{\rightleftharpoons}} E + P$$

Fig. 4.12. Reversible one substrate, one product model with one binary intermediate.

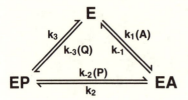

Fig. 4.13. Cyclic representation of a single substrate, two product enzyme-catalyzed reaction.

References

1. Magar, M. E. (1972). *Data Analysis in Biochemistry and Biophysics.*, pp. 20–54, New York, Academic Press.
2. Schulz, A. R. (1991). Algorithms for the derivation of flux and concentration control coefficients. *Biochem. J.* **278**: 299–304.
3. King, E. S. and Altman, C. A. (1956). A systematic method of deriving the rate-laws for enzyme- catalyzed reactions. *J. Phys. Chem.* **60**: 1375–78.
4. Gibbons, A. (1985). *Algorithmic Graph Theory.*, pp. 39–124, Cambridge, Cambridge Univ. Press.
5. Hill, T. L. (1988). Interrelations between random walks on diagrams (graphs) with and without cycles. *Proc. Natl. Acad. Sci. USA* **85**: 2879–83.
6. Hill, T. L. (1988a). Further properties of random walks on diagrams (graphs) with and without cycles. *Proc. Natl. Acad. Sci. USA* **85**: 3271–75.
7. Hill, T. L. (1988b). Number of visits to a state in a random walk before absorption and related topics. *Proc. Natl. Acad. Sci. USA* **85**: 45–81.
8. Hill, T. L. (1988c). Discrete-time random walks on diagrams (graphs) with cycles. *Proc. Natl. Acad. Sci. USA* **85**: 5345–49.
9. Hill, T. L. (1989). *Free Energy Transduction and Biochemical Cycles.*, pp. 39–88, New York, Springer-Verlag.
10. Poland, D. (1989). King-Altman-Hill diagram method for open systems. *J. Phys. Chem.* **93**: 3605–12.
11. Chou, K.-C. (1990). Application of graph theory to enzyme kinetics and protein folding kinetics. Steady state and non-steady state systems. *Biophys. Chem.* **35**: 1–24.

Part Two

Enzyme reaction sequence

5

Multi-reactant enzymic reactions

The enzymic reactions which were considered in the previous section provided for an understanding of the basic mathematical concepts of steady state kinetics. However, there are very few enzymes which catalyze reactions with only one substrate and only one product. The following is a much more representative enzymic reaction,

$$A + B \underset{\text{enzyme}}{\rightleftharpoons} P + Q.$$

The foregoing enzyme-catalyzed reaction is termed a bi-bi reaction. That is, there are two substrates and two products. Most of the pyridine nucleotide dehydrogenases, most of the kinases and most of the aminotransferases are reactions of this type. One way of studying the steady state kinetic behavior of this type of enzymic reaction is to saturate the enzyme with one substrate in the absence of either product and then conduct a substrate-saturation experiment with the other substrate. This procedure could be repeated by switching the roles of the substrates, and if the reaction were reversible, a similar series could be conducted using the products as substrates for the reverse reaction.

5.1 Three distinct two substrate, two product reaction sequences

The kinetic behavior of any enzyme-catalyzed reaction is always simplified by saturation. The difficulty is that much information is lost by such a procedure. Figure 5.1 shows three of many reaction sequences that might be involved in a bi-bi enzymic reaction.

The rate constants in Figure 5.1 contain a double subscript. This is a more explicit symbolism than was employed in the previous section. The enzyme species in the digraphs are numbered in a clockwise manner

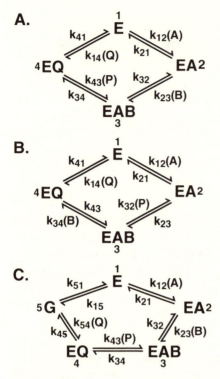

Fig. 5.1. Three models of ordered, two substrate, two product enzyme-catalyzed reactions. A) An ordered, bi-bi, sequential model. B) An ordered, bi-bi, ping-pong model. C) An ordered, bi-bi, iso-sequential model.

starting with the free enzyme. The first subscript identifies the source species and the second subscript of the rate constant identifies the destination species. Figure 5.1A represents an ordered, sequential bi-bi reaction sequence. It is ordered because the binding of the substrates to, and the dissociation of the products from, the enzyme is ordered rather than random. That is, substrate A must bind to the enzyme before substrate B can bind and product P must dissociate before Q. It is a sequential reaction because all of the substrates bind to the enzyme before any of the products dissociate. Logically, the reaction sequence should include an EPQ complex, but the inter-conversion of the EAB and EPQ complexes would not involve the interaction of the enzyme with a reactant, and so, as noted in chapter 4, the omission of the EPQ complex does not affect the rate equation when it is expressed in the kinetic form. Pyridine nucleotide dehydrogenases are examples of the ordered sequential mechanisms, although some of these dehydrogenases are somewhat more complex in that they involve some abortive complexes. The reaction sequence in Figure 5.1B differs from that in Figure 5.1A because the first product dissociates from the enzyme before the binding of the second substrate. This reaction is called an ordered ping-pong bi-bi reaction. Most aminotransferases catalyze this type of reaction sequence. In the case of an amino acid transferase, the enzyme species E would represent the enzyme in the pyridoxal phosphate form and F would represent the enzyme in the pyridoxamine form. Substrate A would be an amino acid while product P would be the corresponding α-keto acid. Substrate B would be the second α-keto acid and Q would be the corresponding amino acid. The reaction sequence portrayed in Figure 5.1C is called an ordered, iso-sequential bi-bi mechanism. This sequence can be envisioned as one in which the enzyme retains the conformation which characterized the EQ complex for a finite period of time after product Q had dissociated. Thus, the free enzyme could exist in either the E or G conformation, and substrate A would bind only to the E species and the product Q would bind only to the G species. If the G species were extremely unstable in an aqueous solution, the conversion of G to E would be essentially irreversible. This is one possible explanation for the kinetic irreversibility of some enzyme-catalyzed reactions when the reaction would be expected to be reversible from thermodynamic considerations.

5.2 The connection matrix method for deriving rate equations

The reader should be apprised of a number of papers which were instrumental in development of a satisfactory treatment of the steady state kinetic behavior of multi-reactant enzymes[1-3]. Graphical implementation of the King-Altman[4] method is widely employed in the derivation of the complete rate equations for multi-reactant enzymic reactions. The graphical method is a great improvement over the repeated inversion of a 5×5 matrix, as would be required in the case of the sequence in Figure 5.1C. However, in the case of highly random reaction sequences, the graphical method also becomes unwieldy. Digraphs such as those shown in Figure 5.1 can be represented as connection matrices. The connection matrix representation of a reaction sequence is more concise than the graphical representation, the analysis of the connection matrix is systematic, and it has the enormous advantage that the algorithm for analysis of the connection matrix can be incorporated easily into a computer program[5,6]. The connection method is based exactly on the same graph theory principles on which the graphical King-Altman method is based. While the connection matrix method was developed for computer-based derivation of rate equations for enzyme-catalyzed reactions, it can be employed easily for manual derivation as well.

The connection matrix associated with the reaction sequence shown in Figure 5.1A is

$$
\begin{vmatrix}
0 & A & 0 & Q \\
1 & 0 & B & 0 \\
0 & 1 & 0 & 1 \\
1 & 0 & P & 0
\end{vmatrix}
$$

The source vertices (species) of the digraph are represented by the rows of the connection matrix, while the destination vertices (species) are represented by the columns of the connection matrix. If there is a directed edge in the digraph from the source vertex to the destination vertex, the element in the row corresponding to the source vertex and the column corresponding to the destination vertex is 1, or if the directed edge is associated with the reaction of the enzyme with a reactant, then the element is that of the reactant. If there is no directed edge in the digraph from the source vertex to the destination vertex, the element is 0. For example, there is no directed edge in Figure 5.1A from node 1 to node 1; therefore the first element in the first row is 0. There is an edge from vertex 1 to vertex 2 and this is associated with the binding of substrate A; therefore the second element in row 1 is A. There is no edge from vertex 1 to vertex 3; therefore the third element in row 1 is 0. There is an edge from vertex 1 to vertex 4 and this is associated with the binding of Q so the fourth element in row 1 is Q. In like manner, there is an edge from vertex 2 to vertex 1 so that the first element in row 2 is 1. There is no edge from vertex 2 to vertex 2 so that the second element in row 2 is 0. The edge from vertex 2 to vertex 3 is associated with the binding of B so that the third element in row 2 is B. The fourth element in row 2 is 0 because there is no edge from vertex 2 to vertex 4. This procedure is continued until the connection matrix has been completed.

The enzyme distribution expression for each enzyme species is obtained by first replacing all of the elements in the row corresponding to the enzyme species in question equal to zero. This provides a means of searching for the spanning trees which terminate at the enzyme node in question. The distribution expression for the free enzyme is obtained as follows.

$$
U = \begin{vmatrix} 0 & 0 & 0 & 0 \\ 1 & 0 & B & 0 \\ 0 & 1 & 0 & 1 \\ 1 & 0 & P & 0 \end{vmatrix}, \quad Q = \begin{vmatrix} 0 & 0 \\ 1 & 3 \\ 2 & 4 \\ 1 & 3 \end{vmatrix}
$$

The connection matrix is U and Q is a secondary matrix which is prepared from U. The Q matrix identifies the non-zero elements in the corresponding row of U. The spanning trees are obtained by an analysis of Q except that the reactants associated with the spanning tree are obtained by reference to U. Each path terminating at the free enzyme is represented as a vector which is constructed by taking one element from each row of Q. One starts with the left-most elements of Q beginning with the first row. As each element is added to the vector, a test is made to determine if a cyclic path is being formed. The test consists of the following rationale.

The position of the element in the vector corresponds to the row of Q from which the element is taken. An index is set to the position of the element in the vector, and this corresponds to the row of Q from which the element has been taken. The test for validity of the path to that point is complete if the value of the element is equal to zero, or if the value of the element is greater than the index, or if the value of the element is equal to the index. If either of the first two conditions is true, then the path is a valid spanning tree to that point, and the index is incremented by one and an element is selected from the next row down in the Q matrix. If the element is equal to the index, then the path is cyclic, and the path is abandoned because it is cyclic. However, if the value of the element is non-zero and less than that of the index, the value of the element is used as a pointer to a previous element in the vector. The element pointed to is subjected to the test described, and this procedure is continued until the path is rejected as a cyclic path or until the path is complete. The path is complete when the vector contains an element from each row of Q. The foregoing procedure is followed until all of the possible paths have been tested and found to be either spanning tress or rejected as cyclic paths.

The algorithm outlined in the previous paragraph can be followed to obtain the enzyme distribution expression for the free enzyme by referring to the previous Q and U matrices. The first element in the vector is taken from the first row of Q and is 0. The partially constructed vector is (0,) and the index is incremented from 1 to 2. The left-most element in the second row of Q is 1. The partially constructed vector is (0, 1,). Since 1 is less than the index, the first element in the vector is pointed to and, since that value is zero, the path is valid to this point and the index is incremented to 3. The left-most element in the third row of Q is 2 and the vector becomes (0, 1, 2,). The third element in the vector is 2 and this points to the second element in the vector which points to the first element which is 0. The path to this point is valid and the index is incremented to 4. The first element in the last row of Q is 1, and that points to 0 so the complete vector is (0, 1, 2, 1). This vector represents a spanning tree. The non-zero elements in the vector can be expressed as rate constants. The first subscript of the rate constant is the source node of the rate constant and that is given by position of the non-zero element in the vector. The second subscript is the element itself. Thus, the vector (0, 1, 2, 1) is equivalent to $k_{21}k_{32}k_{41}$. Following the same procedure, the second par- tially constructed vector is (0, 1, 2,), but instead of taking the first element of the last of Q, the second element of the last row is taken. The complete vector is (0, 1, 2, 3). Since the edge from vertex 4 to vertex 3 is associated with the binding of P to the enzyme, the complete spanning tree is (0, 1, 2, 3)(P) which is equivalent to $k_{21}k_{32}k_{43}$(P). Since the two ele- ments from the last row of Q were employed in the first two spanning trees, the third spanning tree will include the second element of the third row of Q and the first element of the last row of matrix Q. The vector which represents that spanning tree is (0, 1, 4, 1) and it is equivalent to $k_{21}k_{34}k_{41}$. The next possible path is represented by the vector (0, 1, 4, 3). In this vector the index for the last element is 4 which is greater than the value of the element. The last element is a pointer to the third element of the vector and the value of that element is 4 which is equal to the index, so the vector represents a cyclic path and it is rejected. The next path contains the second element of the third row of Q, and the following partial vector is obtained, (0, 3, 2,). The index for the third element is 3 and the element points to the second element whose value is 3. Therefore this vector represents a cyclic path and is rejected. It is obvious that any vector consisting of the second element of the second row and the first element of the third row of Q constitutes a cyclic path. The next vector to be tested consists of (0, 3, 4, 1), and this is a spanning tree which also contains the concentration of B because the edge from node 2 to node 3 of Figure 5.1A involves the binding of B to the enzyme. The vector (0, 3, 4, 1)(B) is equivalent to $k_{23}k_{34}k_{41}$(B). The last possible path is represented by (0, 3, 4, 3), and this is a cyclic path. The numerator of the distribution expression for the free enzyme is summarized as,

$$(0,1,2,1) \doteq k_{21}k_{32}k_{41}$$
$$(0,1,2,3)(P) \doteq k_{21}k_{32}k_{43}(P)$$
$$(0,1,4,1) \doteq k_{21}k_{34}k_{41}$$
$$(0,3,4,1)(B) \doteq k_{23}k_{34}k_{41}(B)$$

The denominator of the distribution expression is the sum of all of the spanning trees for all of the distribution expressions. The foregoing algorithm for obtaining the possible vectors and testing them for validity can be incorporated into a computer program easily.

The following are the U and Q matrices for the distribution expression of the EA complex.

$$U = \begin{vmatrix} 0 & A & 0 & Q \\ 0 & 0 & 0 & 0 \\ 0 & 1 & 0 & 1 \\ 1 & 0 & P & 0 \end{vmatrix}, \quad Q = \begin{vmatrix} 2 & 4 \\ 0 & 0 \\ 2 & 4 \\ 1 & 3 \end{vmatrix}$$

The numerator of the distribution expression comprises the following spanning trees.

$$(2,0,2,1)(A) \doteq k_{12}k_{32}k_{41}(A)$$
$$(2,0,2,3)(A)(P) \doteq k_{12}k_{32}k_{43}(A)(P)$$
$$(2,0,4,1)(A) \doteq k_{12}k_{34}k_{41}(A)$$
$$(4,0,2,3)(P)(Q) \doteq k_{14}k_{34}k_{43}(P)(Q)$$

The matrices required to obtain the numerator of the distribution expression for the EAB complex are the following.

$$U = \begin{vmatrix} 0 & A & 0 & Q \\ 1 & 0 & B & 0 \\ 0 & 0 & 0 & 0 \\ 1 & 0 & P & 0 \end{vmatrix}, \quad Q = \begin{vmatrix} 2 & 4 \\ 1 & 3 \\ 0 & 0 \\ 1 & 3 \end{vmatrix}$$

The spanning trees which constitute the numerator of the distribution expression for the EAB complex are summarized as,

$$(2,3,0,1)(A)(B) \doteq k_{12}k_{23}k_{41}(A)(B)$$
$$(2,3,0,3)(A)(B)(P) \doteq k_{12}k_{23}k_{43}(A)(B)(P)$$
$$(4,1,0,3)(P)(Q) \doteq k_{14}k_{21}k_{43}(P)(Q)$$
$$(4,3,0,3)(B)(P)(Q) \doteq k_{14}k_{23}k_{43}(B)(P)(Q)$$

The following are the matrices associated with the EQ complex.

$$U = \begin{vmatrix} 0 & A & 0 & Q \\ 1 & 0 & B & 0 \\ 0 & 1 & 0 & 1 \\ 0 & 0 & 0 & 0 \end{vmatrix}, \quad Q = \begin{vmatrix} 2 & 4 \\ 1 & 3 \\ 2 & 4 \\ 0 & 0 \end{vmatrix}$$

The numerator of the enzyme distribution expression for the EQ complex is equal to the following spanning trees.

$$(2,3,4,0)(A)(B) \doteq k_{12}k_{23}k_{34}(A)(B)$$
$$(4,1,2,0)(Q) \doteq k_{14}k_{21}k_{32}(Q)$$
$$(4,1,4,0)(Q) \doteq k_{14}k_{21}k_{34}(Q)$$
$$(4,3,4,0)(B)(Q) \doteq k_{14}k_{23}k_{34}(B)(Q)$$

The rate equation for the reaction sequence shown in Fig. 5.1A is

$$v = k_{41}(EQ) - k_{14}(E)(Q) = \left[k_{41}\frac{(EQ)}{E_t} - k_{14}\frac{(E)}{E_t}(Q) \right] E_t \qquad (5.1)$$

The arithmetic indicated in eq. (5.1) is accomplished simply by replacing the zeros in the last elements of the vectors which represent the distribution expression for the EQ complex with ones and replacing the zeros which the first elements in the vectors which represent the distribution expression for E with fours and multiplying these latter vectors by $-1 \times (Q)$. This process is performed as follows.

$$(2,3,4,1)(A)(B) - (4,1,2,1)(Q)$$
$$(4,1,2,1)(Q) - (4,1,2,3)(P)(Q)$$
$$(4,1,4,1)(Q) - (4,1,4,1)(Q)$$
$$(4,3,4,1)(B)(Q) - (4,3,4,1)(B)(Q)$$

The complete rate equation expressed in the convenient vector form is,

$$v = \frac{[(2,3,4,1)(A)(B) - (4,1,2,3)(P)(Q)]E_t}{\begin{array}{ll} (0,1,2,1) & (2,3,4,0)(A)(B) \\ (0,1,4,1) & (2,3,0,1)(A)(B) \\ (2,0,2,1)(A) & (2,0,2,3)(A)(P) \\ (2,0,4,1)(A) & (4,3,4,0)(B)(Q) \\ (0,3,4,1)(B) & (4,1,0,3)(P)(Q) \\ (0,1,2,3)(P) & (4,0,2,3)(P)(Q) \\ (4,1,2,0)(Q) & (2,3,0,3)(A)(B)(P) \\ (4,1,4,0)(Q) & (4,3,0,3)(B)(P)(Q) \end{array}} \qquad (5.2)$$

The same equation would be obtained if the rate of the reaction were expressed in any of the following terms.

$$v = k_{34}(EAB) - k_{43}(EQ)(P) = \left[k_{34}\frac{(EAB)}{E_t} - k_{43}\frac{(EQ)}{E_t}(P) \right]E_t$$

$$v = k_{23}(EA)(B) - k_{32}(EAB) = \left[k_{23}\frac{(EA)}{E_t}(B) - k_{32}\frac{(EAB)}{E_t} \right]E_t$$

$$v = k_{12}(E)(A) - k_{21}(EA) = \left[k_{12}\frac{(E)}{E_t}(A) - k_{21}\frac{(EA)}{E_t} \right]E_t$$

If the steady state approximation is valid, the four expressions are equivalent. Equation (5.2) is a concise and handy manner in which to write the rate equation in the coefficient form. However, eq. (5.2) can be converted easily to its more conventional form.

$$v = \frac{[k_{12}k_{23}k_{34}k_{41}(A)(B) - k_{14}k_{21}k_{32}k_{43}(P)(Q)]E_t}{\begin{aligned} & k_{21}k_{41}(k_{32} + k_{34}) + k_{12}k_{41}(k_{32} + k_{34})(A) \\ & + k_{23}k_{34}k_{41}(B) + k_{21}k_{32}k_{43}(P) + k_{14}k_{21}(k_{32} + k_{34})(Q) \\ & + k_{12}k_{23}(k_{34} + k_{41})(A)(B) + k_{12}k_{32}k_{43}(A)(P) \\ & + k_{14}k_{23}k_{34}(B)(Q) + k_{14}k_{43}(k_{21} + k_{32})(P)(Q) \\ & + k_{12}k_{23}k_{43}(A)(B)(P) + k_{14}k_{23}k_{43}(B)(P)(Q) \end{aligned}} \qquad (5.3)$$

It should be noted that the numerator of the rate equation contains two terms, the first of which is positive and is the product of all the rate constants in the forward direction times the concentration of all of the reactants which bind to the enzyme when the reaction proceeds in the forward direction. The second term is negative and is the product of all of the rate constants in the reverse direction times all the reactants which bind to the enzyme in the reverse direction. This is true of all ordered, reversible reaction sequences. If the reaction sequence is random and reversible, the numerator will contain more than two terms, half of which will be positive and half will be negative. The numerator of the rate equation of an irreversible sequence will contain only positive terms. The denominator of the rate equation contains a total of 16 terms, but these terms are combined to form 11 terms.

Rather than reformulate eqs. (5.2) or (5.3) into the kinetic form at this time, the coefficient form of the rate equations for the sequences in Figure 5.1B and Figure 5.1C will be derived at this point. The rate of the reaction for the sequence in Figure 5.1B is identical to eq. (5.1). The following is the connection matrix for the sequence in Figure 5.1B.

$$U = \begin{vmatrix} 0 & A & 0 & Q \\ 1 & 0 & 1 & 0 \\ 0 & P & 0 & B \\ 1 & 0 & 1 & 0 \end{vmatrix}, \quad Q = \begin{vmatrix} 2 & 4 \\ 1 & 3 \\ 2 & 4 \\ 1 & 3 \end{vmatrix}$$

The enzyme distribution expressions for this sequence are shown in Figure 5.2.

The rate equation in the coefficient vector form is

$$v = \frac{[(2,3,4,1)(A)(B) - (4,1,2,3)(P)(Q)]E_t}{\begin{array}{ll} (2,3,0,1)(A) & (2,3,4,0)(A)(B) \\ (2,3,0,3)(B) & (2,3,0,1)(A)(B) \\ (0,1,4,1)(B) & (2,0,2,1)(A)(P) \\ (0,3,4,1)(B) & (2,0,2,3)(A)(P) \\ (0,1,2,1)(P) & (4,1,4,0)(B)(Q) \\ (0,1,2,3)(P) & (4,3,4,0)(B)(Q) \\ (4,1,0,3)(Q) & (4,1,2,0)(P)(Q) \\ (4,3,0,3)(Q) & (4,0,2,3)(P)(Q) \end{array}}$$

(5.4)

The rate of the reaction shown in Fig. 5.1C is

$$v = k_{51}(G) - k_{15}(E) = \left[k_{51} \frac{(G)}{E_t} - k_{15} \frac{(E)}{E_t} \right] E_t$$

(5.5)

The following is the connection matrix for this reaction sequence.

$$U = \begin{vmatrix} 0 & A & 0 & 0 & 1 \\ 1 & 0 & B & 0 & 0 \\ 0 & 1 & 0 & 1 & 0 \\ 0 & 0 & P & 0 & 1 \\ 1 & 0 & 0 & Q & 0 \end{vmatrix}, \quad Q = \begin{vmatrix} 2 & 5 \\ 1 & 3 \\ 2 & 4 \\ 3 & 5 \\ 1 & 4 \end{vmatrix}$$

$(E)/E_t$	$(EA)/E_t$	$(F)/E_t$	$(EQ)/E_t$
$(0,1,2,1)(P)$	$(2,0,2,1)(A)(P)$	$(2,3,0,1)(A)$	$(2,3,4,0)(A)(B)$
$(0,1,2,3)(P)$	$(2,0,2,3)(A)(P)$	$(2,3,0,3)(A)$	$(4,1,2,0)(P)(Q)$
$(0,1,4,1)(B)$	$(2,0,4,1)(A)(B)$	$(4,1,0,3)(Q)$	$(4,1,4,0)(B)(Q)$
$(0,3,4,1)(B)$	$(4,0,2,3)(P)(Q)$	$(4,3,0,3)(Q)$	$(4,3,4,0)(Q)(B)$

Fig. 5.2. Enzyme distribution expressions in vector form for the ping-pong model shown in Fig. 5.1B.

(E)/E_t	(EA)/E_t	(EAB)/E_t
(0,1,2,3,1)(P)	(2,0,2,3,1)(A)(P)	(2,3,0,2,1)(A)(B)(P)
(0,1,2,3,4)(P)(Q)	(2,0,2,3,4)(A)(P)(Q)	(2,3,0,3,4)(A)(B)(P)(Q)
(0,1,2,5,1)	(2,0,2,5,1)(A)	(2,3,0,5,1)(A)(B)
(0,1,4,5,1)	(2,0,4,5,1)(A)	(5,1,0,3,4)(P)(Q)
(0,3,4,5,1)(B)	(5,0,1,2,4)(P)(Q)	(5,3,0,3,4)(B)(P)(Q)

(EQ)/E_t	(G)/E_t
(2,3,4,0,1)(A)(B)	(2,3,4,5,0)(A)(B)
(2,3,4,0,4)(A)(B)(Q)	(5,1,2,3,0)(P)
(5,1,2,0,4)(Q)	(5,1,2,5,0)
(5,1,4,0,4)(Q)	(5,1,4,5,0)
(5,3,4,0,4)(B)(Q)	(5,3,4,5,0)(B)

Fig. 5.3. Enzyme distribution expressions in vector form for the iso-sequential model shown in Fig. 5.1C.

The enzyme distribution expressions for the reaction sequence portrayed in Figure 5.1C are presented in Figure 5.3.

The rate equation for this reaction sequence in the coefficient form in terms of the vectors is

$$v = \frac{[(2,3,4,5,1)(A)(B) - (5,1,2,3,4)(P)(Q)]E_t}{D}$$

where the denominator D is

$$
\begin{array}{llll}
(0,1,2,5,1) & & & \\
(0,1,4,5,1) & & & \\
(5,1,2,5,0) & (2,3,4,5,0)\,(A)\,(B) & & \\
(5,1,4,5,0) & (2,3,4,0,1)\,(A)\,(B) & (2,3,0,3,1)\,(A)\,(B)\,(P) & \\
 & (2,3,0,5,1)\,(A)\,(B) & & \\
(2,0,2,5,1)\,(A) & & (2,3,4,0,4)\,(A)\,(B)\,(Q) & \\
(2,0,4,5,1)\,(A) & (2,0,2,3,1)\,(A)\,(P) & & \\
 & & (2,0,2,3,4)\,(A)\,(P)\,(Q) & \\
(5,3,4,5,0)\,(B) & (5,3,4,0,4)\,(B)\,(Q) & & \\
(0,3,4,5,1)\,(B) & & (5,3,0,3,4)\,(B)\,(P)\,(Q) & \\
 & (5,1,0,3,4)\,(P)\,(Q) & & \\
(5,1,2,3,0)\,(P) & (5,0,2,3,4)\,(P)\,(Q) & (2,3,0,3,4)\,(A)\,(B)\,(P)\,(Q) & \\
(0,1,2,3,1)\,(P) & (0,1,2,3,4)\,(P)\,(Q) & & \\
(5,1,2,0,4)\,(Q) & & & \\
(5,1,4,0,4)\,(Q) & & &
\end{array}
\tag{5.6}
$$

5.3 Reformulation of the rate equations for multi-reactant enzymes

Equations (5.2), (5.4) and (5.6) are the rate equations in the coefficient form. The task remains to reformulate these equations into the kinetic form such that the equations are expressed in terms of parameters which can be determined in studies of the steady state kinetic behavior of the enzymes.

This task is accomplished by a procedure analogous to that employed in chapter 1. However, the enzymic reactions under consideration contain two substrates and two products. Therefore, each term in the numerator and denominator of the rate equations will be divided by the coefficient of the denominator term which contains all of the substrate concentrations to the highest equal power. This is actually the procedure that was employed in chapter 1. The equation for the maximal velocity in the forward direction in the case of the ordered sequential reaction in Figure 5.1A is

$$
V_f = \frac{\text{num. 1}}{\text{coef. AB}} = \frac{(2,3,4,1)E_t}{\begin{array}{c}(2,3,4,0)\\(2,3,0,1)\end{array}} = \frac{k_{34}k_{41}E_t}{k_{34}+k_{41}}
\tag{5.7}
$$

The maximal velocity in the reverse direction for the sequential model is the second numerator term divided by the coefficient of the denominator term which contains all of the product concentrations to the highest equal power.

$$
V_r = \frac{\text{num. 2}}{\text{coef. PQ}} = \frac{(4,1,2,3)E_t}{\begin{array}{c}(4,1,0,3)\\(4,0,2,3)\end{array}} = \frac{k_{21}k_{32}E_t}{k_{21}+k_{32}}
\tag{5.8}
$$

The Michaelis constants for the substrates are defined as a quotient, the denominator of which is the coefficient of the denominator term which contains the concentrations of all of the substrates to the highest equal power. The numerator of the quotient is the coefficient of the denominator term which contains the concentrations of all the substrates to the highest equal power *except the variable substrate* which it contains to one lower power. For the reaction sequence under consideration, the Michaelis constant for substrate A is the denominator coefficient of (B) divided by the coefficient of the (A)(B) term.

$$
K_a = \frac{\text{coef. B}}{\text{coef. AB}} = \frac{(0,3,4,1)}{\begin{array}{c}(2,3,4,0)\\(2,3,0,1)\end{array}} = \frac{k_{34}k_{41}}{k_{12}(k_{34}+k_{41})}
\tag{5.9}
$$

This definition of the Michaelis constant corresponds exactly with the definition employed in Chapter 1. Only one substrate was involved in the reaction considered in Chapter 1, so the numerator of the Michaelis constant was the constant term since that was the term which contained the concentration of the substrate to the zero power. If the reaction involved three substrates, the Michaelis constant for substrate A would be the coefficient of the denominator term containing BC divided by the coefficient of the denominator term containing (A)(B)(C). Thus, the Michaelis constant for substrate B is,

$$K_b = \frac{\text{coef. A}}{\text{coef. AB}} = \frac{\begin{matrix}(2,0,2,1)\\(2,0,4,1)\end{matrix}}{\begin{matrix}(2,3,4,0)\\(2,3,0,1)\end{matrix}} = \frac{k_{41}(k_{32}+k_{34})}{k_{23}(k_{34}+k_{41})} \tag{5.10}$$

The Michaelis constants for the products are defined in an analogous manner.

$$K_p = \frac{\text{coef. Q}}{\text{coef. PQ}} = \frac{\begin{matrix}(4,1,2,0)\\(4,1,4,0)\end{matrix}}{\begin{matrix}(4,1,0,3)\\(4,0,2,3)\end{matrix}} = \frac{k_{21}(k_{32}+k_{34})}{k_{43}(k_{21}+k_{32})} \tag{5.11}$$

$$K_q = \frac{\text{coef. P}}{\text{coef. PQ}} = \frac{(0,1,2,3)}{\begin{matrix}(4,1,0,3)\\(4,0,2,3)\end{matrix}} = \frac{k_{21}k_{32}}{k_{14}(k_{21}+k_{32})} \tag{5.12}$$

The equilibrium constant is defined,

$$\frac{\text{num. 1}}{\text{num. 2}} = \frac{(2,3,4,1)E_t}{(4,1,2,3)E_t} = \frac{k_{12}k_{23}k_{34}k_{41}}{k_{14}k_{21}k_{32}k_{43}} \tag{5.13}$$

If one considers eqs. (5.2),(5.4) and (5.6), it is apparent that the 16 denominator terms of eq. (5.2) are combined into 11 terms, while the 16 denominator terms of eq. (5.4) are combined into 8 terms and the 25 denominator terms of eq. (5.6) are combined into 14 terms. Since the maximal velocities and Michaelis constants are defined rigidly, it is apparent that some flexibility is going to have to be introduced in the definition of the remaining steady state parameters. Furthermore, it is obvious that the definitions of maximal velocity in the forward direction and the Michaelis constants for the substrates do not acknowledge the existence of products and the definitions of maximal velocity in the reverse direction and the Michaelis constants for the products do not acknowledge the existence of substrates. There are terms in the denominator of all three rate equations which contain the concentrations of both substrates and products. To accommodate this situation, a new class of steady state parameters is defined and, in accordance with the terminology proposed by Cleland[3], these will be called *inhibition constants*. Inhibition constants can be defined in terms of both substrates and products. Like the Michaelis constants, the inhibition constants are defined as quotients of terms from the denominator of the rate equation. The numerator of the quotient will consist of a term which contains the concentration of the variable reactant to one lower power than does the denominator of the quotient. However, there are two terms in the denominator of the rate equation which are *never* used as the denominator of the quotient which defines the inhibition constant. These terms are the coefficient of the term which contains all of the substrates to the highest equal power and the term which contains all of the products to the highest equal power. These two latter terms can be used as the numerator of the quotient, but not as the denominator. To illustrate, an inhibition constant for substrate A could be defined as in eq. (5.12) for the reaction whose rate equation is eq. (5.2).

$$K_{ia} = \frac{\text{constant}}{\text{coef. A}} = \frac{\begin{matrix}(0,1,2,1)\\(0,1,4,1)\end{matrix}}{\begin{matrix}(2,0,2,1)\\(2,0,4,1)\end{matrix}} = \frac{k_{21}}{k_{12}} \tag{5.14}$$

However, for the same rate equation, an inhibition constant for A could also be defined as,

$$K_{ia} = \frac{\text{coef. P}}{\text{coef. AP}} = \frac{(0,1,2,3)}{(2,0,2,3)} = \frac{k_{21}}{k_{12}} \tag{5.15}$$

The foregoing definitions of K_{ia} are equal, but this will not always be the case. An inhibition constant could be defined for substrate B as the constant term in the denominator of eq. (5.2) divided by the coefficient B. To do so would be correct, but it is advantageous to reserve the use of the constant term to define *inhibition constants* of those reactants which bind to the free enzyme. Thus, for the reaction sequence portrayed in Figure 5.1A, the constant term will be used in the definition of inhibition constants for substrate A and product Q.

$$K_{iq} = \frac{\text{constant}}{\text{coef. Q}} = \frac{\text{coef. B}}{\text{coef. BQ}} = \frac{k_{41}}{k_{14}} \tag{5.16}$$

In the case of eq. (5.2), there are two definitions of inhibition constants for A and also Q which are equal. However, not all multiple definitions of inhibition constants are equal. The following inhibition constants for P can be defined for eq. (5.2).

$$K_{ip_1} = \frac{\text{coef. A}}{\text{coef. AP}} = \frac{\begin{matrix}(2,0,2,1)\\(2,0,4,1)\end{matrix}}{(2,0,2,3)} = \frac{k_{41}(k_{32}+k_{41})}{k_{32}k_{43}} \tag{5.17}$$

$$K_{ip_2} = \frac{\text{coef. AB}}{\text{coef. ABP}} = \frac{\begin{matrix}(2,3,4,0)\\(2,3,0,1)\end{matrix}}{(2,3,0,3)} = \frac{(k_{34}+k_{41})}{k_{43}} \tag{5.18}$$

$$v = \cfrac{\left(\dfrac{\text{num. 1}}{\text{coef. AB}}\right)(A)(B) - \left(\dfrac{\text{num. 1}}{\text{coef. AB}} \times \dfrac{\text{num. 2}}{\text{num. 1}}\right)(P)(Q)}{\begin{aligned} &\left(\dfrac{\text{constant}}{\text{coef. A}} \times \dfrac{\text{coef. A}}{\text{coef. AB}}\right) + \left(\dfrac{\text{coef. A}}{\text{coef. AB}}\right)(A) + \left(\dfrac{\text{coef. B}}{\text{coef. AB}}\right)(B) \\ &+ \left(\dfrac{\text{coef. AB}}{\text{coef. AB}}\right)(A)(B) + \left(\dfrac{\text{coef. P}}{\text{coef. AP}} \times \dfrac{\text{coef. AP}}{\text{coef. A}} \times \dfrac{\text{coef. A}}{\text{coef. AB}}\right)(P) \\ &+ \left(\dfrac{\text{coef. Q}}{\text{constant}} \times \dfrac{\text{constant}}{\text{coef. A}} \times \dfrac{\text{coef. A}}{\text{coef. AB}}\right)(Q) \\ &+ \left(\dfrac{\text{coef. AP}}{\text{coef. A}} \times \dfrac{\text{coef. A}}{\text{coef. AB}}\right)(A)(P) + \left(\dfrac{\text{coef. BQ}}{\text{coef. B}} \times \dfrac{\text{coef. B}}{\text{coef. AB}}\right)(B)(Q) \\ &+ \left(\dfrac{\text{coef. PQ}}{\text{coef. Q}} \times \dfrac{\text{coef. Q}}{\text{constant}} \times \dfrac{\text{constant}}{\text{coef. A}} \times \dfrac{\text{coef. A}}{\text{coef. AB}}\right)(P)(Q) \\ &+ \left(\dfrac{\text{coef. ABP}}{\text{coef. AB}}\right)(A)(B)(P) \\ &+ \left(\dfrac{\text{coef. BPQ}}{\text{coef. BQ}} \times \dfrac{\text{coef. BQ}}{\text{coef. B}} \times \dfrac{\text{coef. B}}{\text{coef. AB}}\right)(B)(P)(Q) \end{aligned}}$$

Fig. 5.4 Reformulation of the rate equation for an ordered, bi-bi, sequential enzyme model from the coefficient to the kinetic form.

$$K_{ip_3} = \frac{\text{coef. BQ}}{\text{coef. BPQ}} = \frac{(4,3,4,0)}{(4,3,0,3)} = \frac{k_{34}}{k_{43}} \tag{5.19}$$

With the foregoing definitions of steady state parameters, eq. (5.2) can be reformulated into the kinetic form. The process is shown in Figure 5.4. The rate equation for the ordered sequential reaction shown in Figure 5.1A is

$$v = \cfrac{V_f\left[(A)(B) - \dfrac{(P)(Q)}{K_{eq}}\right]}{\begin{aligned} &K_{ia}K_b + K_b(A) + K_a(B) + (A)(B) \\ &+ \dfrac{K_{ia}K_b}{K_{ip_1}}(P) + \dfrac{K_{ia}K_b}{K_{iq}}(Q) + \dfrac{K_b}{K_{ip_1}}(A)(P) + \dfrac{K_a}{K_{iq}}(B)(Q) \\ &+ \dfrac{K_{ia}K_b}{K_p K_{iq}}(P)(Q) + \dfrac{(A)(B)(P)}{K_{ip_2}} + \dfrac{K_a}{K_{ip_3}K_{iq}}(B)(P)(Q) \end{aligned}} \tag{5.20}$$

The maximal velocities and Michaelis constants are defined in a rigid manner, but there is flexibility in the definition of the inhibition constants.

For this reason, different individuals may reformulate the rate equation somewhat differently. It is desirable for the reformulated equation to appear as tidy as possible. The denominator of eq. (5.20) contains terms which include a product concentration divided by an inhibition constant for that product. This is analogous to the equations derived in chapter 3 for inhibition. This is in contrast to K_{ia}, the inhibition constant for substrate A. In the latter case, the inhibition constant functions in a manner similar to a Michaelis constant. That is, it is multiplied by the concentration of B. However, if the rate equation were derived for the reverse reaction, A and B would be the products and the denominator of the rate equation would include terms in which the concentration of A would be divided by K_{ia}. This is the logic behind the terminology used.

Equation (5.4) can be reformulated into the kinetic form by following the procedure detailed for eq. (5.20).

$$
v = \frac{V_f \left[(A)(B) - \dfrac{(P)(Q)}{K_{eq}} \right]}{\begin{array}{l} K_b(A) + K_a(B) + (A)(B) + \dfrac{K_{ia}K_b}{K_{ip}}(P) + \dfrac{K_a K_{ib}}{K_{iq}}(Q) \\[3mm] + \dfrac{K_b}{K_{ip}}(A)(P) + \dfrac{K_a}{K_{iq}}(B)(Q) + \dfrac{K_a K_{ib}}{K_p K_{iq}}(P)(Q) \end{array}}
\tag{5.21}
$$

In like manner the complete rate equation in the kinetic form can be obtained for the iso-sequential reaction sequence portrayed in Figure 5.1C.

$$
v = \frac{V_f \left[(A)(B) - \dfrac{(P)(Q)}{K_{eq}} \right]}{\begin{array}{l} K_{ia}K_b + K_b(A) + K_a(B) + (A)(B) \\[3mm] + \dfrac{K_{ia}K_b}{K_{ip_1}}(P) + \dfrac{K_{ia}K_b}{K_{iq_1}}(Q) + \dfrac{K_b}{K_{ip_1}}(A)(P) \\[3mm] + \dfrac{K_a}{K_{iq_1}}(B)(Q) + \dfrac{K_{ia}K_b}{K_p K_{iq_1}}(P)(Q) + \dfrac{(A)(B)(P)}{K_{ip_2}} \\[3mm] + \dfrac{(A)(B)(Q)}{K_{iq_2}} + \dfrac{K_b}{K_{ip_1}K_{iq_3}}(A)(P)(Q) \\[3mm] + \dfrac{K_a}{K_{ip_3}K_{iq_1}}(B)(P)(Q) + \dfrac{(A)(B)(P)(Q)}{K_{ip_3}K_{iq_2}} \end{array}}
\tag{5.22}
$$

The complete rate equations in the kinetic form for the reaction sequences portrayed in Figure 5.1 contain a great deal of useful information which can be deduced from studies of the steady state kinetic behavior of the enzymes. The analysis of this information will be discussed in the following chapter.

5.4 Problems for chapter 5

5.1   An enzyme catalyzes the following reaction

$$A + B + C \rightleftharpoons P + Q.$$

The reaction sequence for the enzyme-catalyzed reaction is shown in Figure 5.5.

a) Derive the enzyme distribution expressions for the foregoing reaction sequence and write the complete rate equation in the vector coefficient form.

b) Write the expressions for the steady state parameters for the reaction sequence.

c) Reformulate the rate equation into the kinetic form. (Note: retain the rate equation in the kinetic form for it will be required to solve problems in chapter 6.)

5.2   An enzyme catalyzes the following reaction

$$A + B \rightleftharpoons P + Q + R.$$

The reaction sequence of the enzyme-catalyzed reaction is shown in Figure 5.6

a) Derive the enzyme distribution expressions for the foregoing reaction sequence and write the complete rate equation in the vector coefficient form.

b) Write the expressions for the steady state parameters for the reaction sequence.

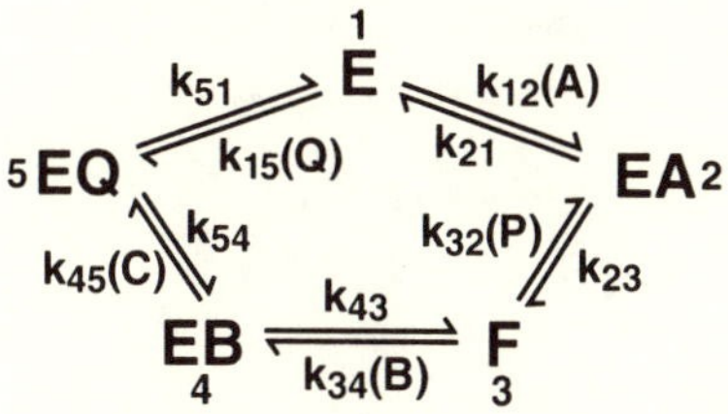

Fig. 5.5. Model of an ordered, ter-bi, ping-pong enzyme-catalyzed reaction.

Multi-reactant enzymic reactions

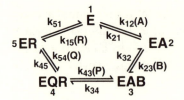

Fig. 5.6 Model of an ordered, bi-ter, sequential enzyme-catalyzed reaction.

c) Reformulate the rate equation into the kinetic form. (Note: retain the rate equation in the kinetic form for it will be required to solve problems in chapter 6.)

References

1. Segal, H. L., Kachmar, J. F. and Boyer, P. D. (1951). Kinetic analysis of enzyme reactions. *Enzymologia* **15**: 187–98.
2. Wong, J. T-F. and Hanes, C. S. (1962). Kinetic formulations for enzymic reactions involving two substrates. *Canad. J. Biochem. Physiol.* **40**: 763–804.
3. Cleland, W. W. (1963). The kinetics of enzyme-catalyzed reactions with two or more substrates or products. I. Nomenclature and rate equations. *Biochim. Biophys. Acta* **67**: 104–37.
4. King, E. S. and Altman, C. A. (1956). A systematic method of deriving the rate-laws for enzyme-catalyzed reactions. *J. Phys. Chem.* **60**: 1375–78.
5. Fisher, D. D. and Schulz, A. R. (1969). Connection matrix representation of enzyme reaction sequences. *Math. Biosci.* **4**: 189–200.
6. Fisher, D. D. and Schulz, A. R. (1970). Computer-based derivation and reformulation of enzyme reaction models. *Biomed. Comp.* **1**: 221–35.

6

Analysis of multi-reactant enzyme kinetics

The rate equations for an ordered sequential, an ordered ping-pong and an ordered iso-sequential reactions sequences were derived in chapter 5. These equations contain much information, and in this chapter the analysis of this information will be analysed. The reader should be aware of a number of excellent references to this type of analysis[1-4].

6.1 Analysis of the kinetic behavior of an enzymic reaction in the absence of products

The ordered sequential reaction sequence will be considered first, and for this purpose it is convenient to rearrange eq. (5.18) by dividing each term on the right-hand side of the equation by the concentrations of the substrates.

$$v = \cfrac{V_f\left[1 - \cfrac{\Gamma}{K_{eq}}\right]}{\begin{aligned} &1 + \frac{K_a}{(A)} + \frac{K_b}{(B)} + \frac{K_{ia}K_b}{(A)(B)} + \frac{(P)}{K_{ip_2}} \\ &+ \frac{K_b(P)}{K_{ip_1}(B)} + \frac{K_{ia}K_b(P)}{K_{ip_1}(A)(B)} + \frac{K_a(Q)}{K_{iq}(A)} \\ &+ \frac{K_{ia}K_b(Q)}{K_{iq}(A)(B)} + \frac{K_a(P)(Q)}{K_{ip_3}K_{iq}(A)} + \frac{K_{ia}K_b(P)(Q)}{K_pK_{iq}(A)(B)} \end{aligned}} \tag{6.1}$$

In eq. (6.1), Γ is the mass action ratio, that is, it is the ratio $(P)(Q)/(A)(B)$. If the concentrations of both products are set equal to zero, eq. (6.1) becomes

$$v = \cfrac{V_f}{1 + \cfrac{K_a}{(A)} + \cfrac{K_b}{(B)} + \cfrac{K_{ia}K_b}{(A)(B)}} \tag{6.2}$$

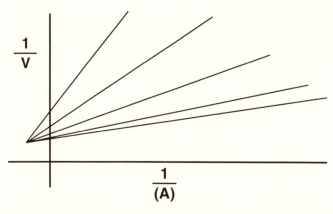

Fig. 6.1. Lineweaver-Burk plots of an ordered, bi-bi, sequential enzyme reaction sequence when A is the variable substrate and B is the non-varied substrate. Each line represents a substrate-saturation curve at a different concentration of B. As the concentration of B increases, the slope and the intercepts of the plot decrease until the enzyme is saturated with B.

The following is the Lineweaver-Burk equation under the condition that $(P) = (Q) = 0$ where A is the variable substrate and B is held constant, that is, B is the non-varied substrate.

$$\frac{1}{v_a} = \frac{K_a}{V_f}\left[1 + \frac{K_{ia}K_b}{K_a(B)}\right]\frac{1}{(A)} + \frac{1}{V_f}\left[1 + \frac{K_b}{(B)}\right] \tag{6.3}$$

It is obvious that if a series of substrate saturation experiments were conducted at different concentrations of B, that both the slope and intercepts of the Lineweaver-Burk plots would decrease with increasing concentrations of B until the enzyme was saturated with B. When the enzyme is saturated with B, $K_b/(B) = 0$. Figure 6.1 portrays the Lineweaver-Burk plots of such a series of experiments. The apparent maximal velocity for any of the lines of Figure 6.1 is given by eq. (6.4).

$$V_f^{app} = \frac{V_f}{1 + \frac{K_b}{(B)}} \tag{6.4}$$

The apparent maximal velocity will increase with increasing concentrations of the non-varied substrate until the enzyme is saturated with the non-varied substrate. The situation with the apparent Michaelis constant

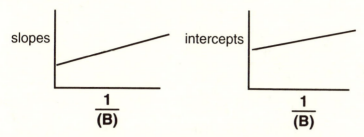

Fig. 6.2. Secondary plots of the slopes and intercepts of the Lineweaver-Burk plots in Figure 6.1 as functions of the reciprocal concentrations of B.

is not as predictable as shown by eq. (6.5).

$$K_a^{app} = K_a \left[\frac{1 + \dfrac{K_{ia}K_b}{K_a(B)}}{1 + \dfrac{K_b}{(B)}} \right] \tag{6.5}$$

Thus, the apparent Michaelis constant may increase, decrease or remain constant with increasing concentrations of B. Estimates of V_f, K_a, K_{ia} and K_b can obtained from the following linear relationships between the slopes and intercepts of the Lineweaver-Burk plots and $1/(B)$.

$$\text{slopes} = \frac{K_{ia}K_b}{V_f} \frac{1}{(B)} + \frac{K_a}{V_f} \tag{6.6}$$

$$\text{intercepts} = \frac{K_b}{V_f(B)} \frac{1}{} + \frac{1}{V_f} \tag{6.7}$$

The two secondary plots provide for the estimation of the parameters. The secondary plots shown in Figure 6.2 are subject to the same criticism that was made in Chapter 1 of the use of the Lineweaver-Burk plots to obtain quantitative estimates of steady state parameters. This can be improved by simply multiplying both sides of eqs. (6.6) and (6.7) by the concentration of B and plotting slope x (B) and intercept x (B) against the concentration of B.

Rather than continue with the analysis of eq. (6.1) at this point, an analysis of the equations for the ping-pong and iso-sequential equations will be considered for the condition where the concentrations of both products are equal to zero. Rearrangement of eq. (5.19) by dividing all of the terms in the numerator and denominator by the concentrations of A and B gives,

$$v = \cfrac{V_f\left[1 - \cfrac{\Gamma}{K_{eq}}\right]}{\begin{aligned} &1 + \frac{K_a}{(A)} + \frac{K_b}{(B)} + \frac{K_b(P)}{K_{ip}(B)} + \frac{K_{ia}K_b(P)}{K_{ip}(A)(B)} \\ &+ \frac{K_a(Q)}{K_{iq}(A)} + \frac{K_aK_{ib}(Q)}{K_{iq}(A)(B)} + \frac{K_aK_{ib}(P)(Q)}{K_pK_{iq}(A)(B)} \end{aligned}} \tag{6.8}$$

If $(P) = (Q) = 0$, eq. (6.8) becomes,

$$v = \cfrac{V_f}{1 + \cfrac{K_a}{(A)} + \cfrac{K_b}{(B)}} \tag{6.9}$$

Equation (6.10) is the Lineweaver-Burk equation for the ping-pong model if A is the variable substrate and B is the non-varied substrate.

$$\frac{1}{v_a} = \frac{K_a}{V_f}\frac{1}{(A)} + \frac{1}{V_f}\left[1 + \frac{K_b}{(B)}\right] \tag{6.10}$$

It is obvious that the non-varied substrate does not affect the slope of the Lineweaver-Burk plots of a ping-pong mechanism, but that the intercepts decrease with increasing concentrations of the non-varied substrate. Thus, it is possible to distinguish between the sequential and ping-pong reaction sequences by conducting a series of substrate-saturation experiments at

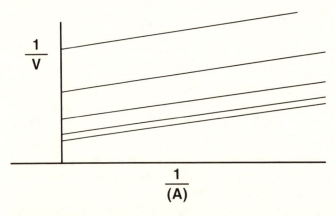

Fig. 6.3. Lineweaver-Burk plots of an ordered, ping-pong enzyme reaction sequence when A is the variable substrate and B is the non-varied substrate. As the concentration of B increases the intercepts of the plots decrease until the enzyme is saturated with B.

different concentrations of the non-varied substrate. This distinction is not apparent if a single substrate-saturation experiment is conducted at a saturating concentration of the non-varied substrate because, in that case, the Michaelis and/or inhibition constant divided by the concentration of the non-varied substrate is equal to zero, and eqs. (6.2) and (6.9) both degrade to eq. (6.11).

$$v = \frac{V_f}{1 + \dfrac{K_a}{(A)}} \tag{6.11}$$

The apparent maximal velocity in the forward direction for the ping-pong reaction sequence is

$$V_f^{app} = \frac{V_f}{1 + \dfrac{K_b}{(B)}} \tag{6.12}$$

The apparent Michaelis constant for the variable substrate is,

$$K_a^{app} = \frac{K_a}{1 + \dfrac{K_b}{(B)}} \tag{6.13}$$

Both the apparent maximal velocity and the Michaelis constants increase to the same extent with increasing concentrations of the non-varied substrate until the enzyme is saturated with the non-varied substrate. This is in contrast to the behavior of the ordered sequential reaction sequence.

Equation (5.20) for the iso-sequential model can be rearranged as eq. (6.14).

$$v = \frac{V_f\left[1 - \dfrac{\Gamma}{K_{eq}}\right]}{\begin{aligned} &1 + \frac{K_a}{(A)} + \frac{K_b}{(B)} + \frac{K_{ia}K_b}{(A)(B)} + \frac{(P)}{K_{ip_2}} \\ &+ \frac{K_b(P)}{K_{ip_1}(B)} + \frac{K_{ia}K_b(P)}{K_{ip_1}(A)(B)} + \frac{(Q)}{K_{iq_2}} \\ &+ \frac{K_a(Q)}{K_{iq_1}(A)} + \frac{K_{ia}K_b(Q)}{K_{iq_1}(A)(B)} + \frac{(P)(Q)}{K_{ip_3}K_{iq_2}} \\ &+ \frac{K_a(P)(Q)}{K_{ip_3}K_{iq_1}(A)} + \frac{K_b(P)(Q)}{K_{ip_1}K_{iq_3}} + \frac{K_{ia}K_b(P)(Q)}{K_p K_{iq_1}(A)(B)} \end{aligned}} \tag{6.14}$$

If the concentrations of both products are set to zero, the resulting equation for the rate of the reaction is identical to eq. (6.2). This means that while one can distinguish between the ping-pong reaction sequence, on the one hand, and the sequential or the iso-sequential on the other, one cannot distinguish between the sequential and the iso-sequential reaction sequences under the condition that the concentrations of both products are equal to zero. The problem lies in the fact that, if the concentrations of both products are equal to zero, only 4 of the 11 denominator terms in eq. (6.1) and only four of the 14 denominator terms in eq. (6.14) exert an effect on the kinetic behavior of the reaction. Conditions must be sought whereby additional denominator terms exert an effect on the kinetic behavior of the reaction if one is to distinguish between these two mechanisms.

6.2 Product inhibition as a tool in the analysis of reaction sequences

Product inhibition provides a means by which more of the denominator terms in these equations exert an influence on the kinetic behavior of the reaction. If either of the product concentrations were *not* equal to zero, while the concentration of the other product were equal to zero, the second numerator term would remain equal to zero. That is, the reaction would remain infinitely far from equilibrium, but additional denominator terms would influence the kinetics of the reaction. Product inhibition is distinct from the general type of inhibition discussed in chapter 3 because a product of an enzymic reaction is a normal reactant in the catalytic cycle. The power of product inhibition resides in this fact.

In the case of the sequential reactions sequence, if the reaction were conducted under conditions where P were present initially, but Q were absent, eq. (6.1) would become

$$
v = \frac{V_f}{1 + \dfrac{K_a}{(A)} + \dfrac{K_b}{(B)} + \dfrac{K_{ia}K_b}{(A)(B)} + \dfrac{(P)}{K_{ip_2}} + \dfrac{K_b(P)}{K_{ip_1}(B)} + \dfrac{K_{ia}K_b(P)}{K_{ip_1}(A)(B)}}
\tag{6.15}
$$

Equation (6.15) contains seven denominator terms in contrast to eq. (6.2) which contains only four denominator terms. When A is the variable substrate, the Lineweaver-Burk equation is

$$
\frac{1}{v_a} = \frac{K_a}{V_f}\left[1 + \frac{K_{ia}K_b}{K_a(B)}\left(1 + \frac{(P)}{K_{ip_1}}\right)\right]\frac{1}{(A)}
$$
$$
+ \frac{1}{V_f}\left[1 + \frac{K_b}{(B)}\left(1 + \frac{(P)}{K_{ip_1}}\right) + \frac{(P)}{K_{ip_2}}\right]
\tag{6.16}
$$

An inspection of eq. (6.16) shows that P will affect the slope of the Lineweaver-Burk plot *provided* the concentration of B, the non-varied substrate, is less than saturating. The presence of P will also affect the intercept of the Lineweaver-Burk plot. The effect of P on the slope will probably differ from its effect on the intercept, but both will be affected and thus, under these conditions, P will serve as a *mixed type inhibitor*. However, if the enzyme were saturated with B such that $K_b/(B) = 0$, the slope would no longer be affected by P. The effect of P on the intercept would decrease, but the intercept would still be affected by the term $(P)/K_{ip2}$. Under these conditions, the product P would function as an uncompetitive inhibitor. The experiments which are necessary are that, when A is the variable substrate, a series of substrate-saturation experiments be conducted each at one of several concentrations of P and this series of experiments should be conducted where the concentration of B is less than saturating, and a second series of experiments where the concentration of B is sufficient to saturate the enzyme. Thus, if five different concentrations of P were employed, this would require a total of ten substrate-saturation experiments.

Equation (6.17) is the Lineweaver-Burk equation for the situation where $(P) = 0$ and $(Q) = 0$ and B is the variable substrate.

$$\frac{1}{v_b} = \frac{K_b}{V_f}\left[1 + \frac{(P)}{K_{ip_1}} + \frac{K_{ia}}{(A)}\left(1 + \frac{(P)}{K_{ip_1}}\right)\right]\frac{1}{(B)} + \frac{1}{V_f}\left[1 + \frac{K_a}{(A)} + \frac{(P)}{K_{ip_2}}\right] \quad (6.17)$$

At a sub-saturating concentration of A, the non-varied substrate, P affects the slope of the Lineweaver-Burk plot and it also affects the intercept. Once again, the effect is not the same on the slope and the intercept, but both are affected and thus P is a mixed type inhibitor. If the enzyme were saturated with A, the effect of P on the slope would be diminished because $K_{ia}/(A) = 0$, but P would still affect the slope and the intercept, and so P would continue to function as a mixed type inhibitor. Thus, the effect of P on kinetic behavior is different when A is the variable substrate than when B is the variable substrate.

Additional information can be obtained when Q, rather than P, is employed as the product inhibitor. Under this condition, the rate of the reaction is given by eq. (6.18).

$$v = \frac{V_f}{1 + \dfrac{K_a}{(A)} + \dfrac{K_b}{(B)} + \dfrac{K_{ia}K_b}{(A)(B)} + \dfrac{K_a(Q)}{K_{iq}(A)} + \dfrac{K_{ia}K_b(Q)}{K_{iq}(A)(B)}} \quad (6.18)$$

When A is the variable substrate and B is the non-varied substrate, the Lineweaver-Burk equation is

$$\frac{1}{v_a} = \frac{K_a}{V_f}\left[1 + \frac{(Q)}{K_{iq}} + \frac{K_{ia}K_b}{K_a(B)}\left(1 + \frac{(Q)}{K_{iq}}\right)\right]\frac{1}{(A)} + \frac{1}{V_f}\left[1 + \frac{K_b}{(B)}\right] \qquad (6.19)$$

It is obvious that Q affects the slope but not the intercept of the Lineweaver-Burk plot and this is true whether or not the enzyme is saturated with the non-varied substrate. Thus, Q is a competitive inhibitor of substrate A. Equation (6.20) is the Lineweaver-Burk equation for the sequential mechanism when Q is the inhibitor and B is the variable substrate.

$$\frac{1}{v_b} = \frac{K_b}{V_f}\left[1 + \frac{K_{ia}}{(A)}\left(1 + \frac{(Q)}{K_{iq}}\right)\right]\frac{1}{(B)} + \frac{1}{V_f}\left[1 + \frac{K_a}{(A)}\left(1 + \frac{(Q)}{K_{iq}}\right)\right] \qquad (6.20)$$

At a sub-saturating concentration of A, the product Q affects both the slope and the intercept of the Lineweaver-Burk plot and, therefore, is a mixed type inhibitor. However, saturation of the enzyme with A eliminates the effect on both the slope and the intercept of the plot. Hence, when the enzyme is saturated with A, the product Q has no inhibitory effect. The concentration of A required to saturate the enzyme in the presence of Q will be greater than that required to saturate the enzyme in the absence of Q, but for any given concentration of Q, once the enzyme is saturated with A, Q is no longer an inhibitor. This observation should be anticipated by inspection of the reaction sequence for the ordered, sequential model portrayed in Figure 5.1A. In that model both A and Q interact with the same species of the enzyme, namely, the free enzyme. This can also be predicted by an inspection of the enzyme distribution expressions for the sequential enzyme model listed in chapter 5. Of the distribution expressions for the sequential reaction sequence, the concentration of A is completely absent only from the distribution expression for the free enzyme. Thus, when the concentration of A approaches infinity, the fraction of enzyme present as E approaches zero.

On the basis of the foregoing discussion, it is possible to construct a product inhibition pattern for the ordered, sequential enzyme model portrayed in Figure 5.1A. The product inhibition pattern for this model is shown in Figure 6.4.

A similar analysis of the product inhibition pattern can be made for the iso-sequential enzyme model shown in Figure 5.1C. The rate equation for this reaction sequence, when P is the product inhibitor but the concentration of Q is equal to zero, is identical to eq. (6.15). Hence, the product

(A)	(B)	Inhibitor	Type of inhibition
Variable	Subsaturate	P	Mixed type
Variable	Saturate	P	Uncompetitive
Subsaturate	Variable	P	Mixed type
Saturate	Variable	P	Mixed type
Variable	Subsaturate	Q	Competitive
Variable	Saturate	Q	Competitive
Subsaturate	Variable	Q	Mixed type
Saturate	Variable	Q	No inhibition

Fig. 6.4. Product inhibition pattern for the sequential enzyme reaction sequence portrayed in Figure 5.1A.

inhibition pattern for this reaction sequence is identical to the ordered, sequential sequence when P is the product inhibitor. Equation (6.21) is the rate equation for the ordered, iso-sequential sequence when Q is the product inhibitor.

$$v = \cfrac{V_f}{1 + \dfrac{K_a}{(A)} + \dfrac{K_b}{(B)} + \dfrac{K_{ia}K_b}{(A)(B)} + \dfrac{(Q)}{K_{iq_2}} + \dfrac{K_a(Q)}{K_{iq_1}(A)} + \dfrac{K_{ia}K_b(Q)}{K_{iq_1}(A)(B)}} \tag{6.21}$$

The Lineweaver-Burk equation for this situation where A is the variable substrate is

$$\frac{1}{v_a} = \frac{K_a}{V_f}\left[1 + \frac{(Q)}{K_{iq_1}} + \frac{K_{ia}K_b}{K_a(B)}\left(1 + \frac{(Q)}{K_{iq_1}}\right)\right]\frac{1}{(A)} + \frac{1}{V_f}\left[1 + \frac{K_b}{(B)} + \frac{(Q)}{K_{iq_2}}\right] \tag{6.22}$$

It can be seen that, when A is the variable substrate, Q will function as a mixed type inhibitor regardless of whether or not the enzyme is saturated with the non-varied substrate. This is in sharp contrast to the sequential reaction sequence where Q functioned as a competitive inhibitor. The distinction between the sequential and the iso-sequential is also seen in the Lineweaver-Burk equation when B is the variable substrate.

$$\frac{1}{v_b} = \frac{K_b}{V_f}\left[1 + \frac{K_{ia}}{(A)}\left(1 + \frac{(Q)}{K_{iq_1}}\right)\right]\frac{1}{(B)} + \frac{1}{V_f}\left[1 + \frac{K_a}{(A)}\left(1 + \frac{(Q)}{K_{iq_1}}\right) + \frac{(Q)}{K_{iq_2}}\right] \tag{6.23}$$

When B is the variable substrate, Q affects both the slope and the intercept of the Lineweaver-Burk plot when A, the non-varied substrate, is present at a subsaturating concentration. Thus, it is a mixed type inhibitor. However, when the enzyme is saturated with A, Q does not affect the slope and, therefore, Q is an uncompetitive inhibitor. The product inhibition pattern for the iso-sequential reaction sequence is summarized in Figure 6.5.

(A)	(B)	Inhibitor	Type of inhibition
Variable	Subsaturate	P	Mixed type
Variable	Saturate	P	Uncompetitive
Subsaturate	Variable	P	Mixed type
Saturate	Variable	P	Mixed type
Variable	Subsaturate	Q	Mixed type
Variable	Saturate	Q	Mixed type
Subsaturate	Variable	Q	Mixed type
Saturate	Variable	Q	Uncompetitive

Fig. 6.5. Product inhibition pattern for the iso-sequential enzyme reaction sequence shown in Figure 5.1C.

Product inhibition provides a means of distinguishing between the sequential and the iso-sequential reaction sequences by exploiting the differences in the complete rate equations.

While it is possible to distinguish between the ping-pong sequence of Figure 5.1B and either of the other two sequences in Figure 5.1 without resorting to product inhibition, it is possible to envision reaction sequences which would not be distinct from the ping-pong model without employing product inhibition. For this reason, the product inhibition pattern for the model in Figure 5.1B will be developed. Equation (6.24) is a rearranged form of the rate equation for this model in which the terms containing (P)(Q) are omitted

$$v = \frac{V_f}{1 + \dfrac{K_a}{(A)} + \dfrac{K_b}{(B)} + \dfrac{K_b(P)}{K_{ip}(B)} + \dfrac{K_{ia}K_b(P)}{K_{ip}(A)(B)} + \dfrac{K_a(Q)}{K_{iq}(A)} + \dfrac{K_aK_{ib}(Q)}{K_{iq}(A)(B)}} \qquad (6.24)$$

The following is the Lineweaver-Burk equation when A is the variable substrate and P is the product inhibitor.

$$\frac{1}{v_a} = \frac{K_a}{V_f}\left[1 + \frac{K_{ia}K_b(P)}{K_aK_{ip}(B)}\right]\frac{1}{(A)} + \frac{1}{V_f}\left[1 + \frac{K_b}{(B)}\left(1 + \frac{(P)}{K_{ip}}\right)\right] \qquad (6.25)$$

At a subsaturating concentration of B, the product P serves as a mixed type inhibitor, but when the enzyme is saturated with B, the product P affects neither the slope nor the intercept and no inhibition would be observed. The Lineweaver-Burk equation when B is the variable substrate and P is the product inhibitor is given by eq. (6.26).

$$\frac{1}{v_b} = \frac{K_b}{V_f}\left[1 + \frac{(P)}{K_{ip}} + \frac{K_{ia}(P)}{K_{ip}(A)}\right]\frac{1}{(B)} + \frac{1}{V_f}\left[1 + \frac{K_a}{(A)}\right] \qquad (6.26)$$

(A)	(B)	Inhibitor	Type of inhibition
Variable	Subsaturate	P	Mixed type
Variable	Saturate	P	No inhibition
Subsaturate	Variable	P	Competitive
Saturate	Variable	P	Competitive
Variable	Subsaturate	Q	Competitive
Variable	Saturate	Q	Competitive
Subsaturate	Variable	Q	Mixed type
Saturate	Variable	Q	No inhibition

Fig. 6.6. Product inhibition pattern for the ping-pong enzyme reaction sequence portrayed in Figure 5.1B.

It is obvious that P affects only the slope of eq. (6.26), and this is true regardless of whether or not the enzyme is saturated with the non-varied substrate. Thus, P is a competitive inhibitor when B is the variable substrate. When A is the variable substrate and Q is the product inhibitor, the Lineweaver-Burk equation is

$$\frac{1}{v_a} = \frac{K_a}{V_f}\left[1 + \frac{(Q)}{K_{iq}} + \frac{K_{ib}(Q)}{K_{iq}(B)}\right]\frac{1}{(A)} + \frac{1}{V_f}\left[1 + \frac{K_b}{(B)}\right] \tag{6.27}$$

It can be determined readily from eq. (6.27) that product Q is a competitive inhibitor when A is the variable substrate regardless of the concentration of the non-varied substrate. Equation (6.28) is the Lineweaver-Burk equation when B is the variable substrate and Q is the product inhibitor.

$$\frac{1}{v_b} = \frac{K_b}{V_f}\left[1 + \frac{K_a K_{ib}(Q)}{K_b K_{iq}(A)}\right]\frac{1}{(B)} + \frac{1}{V_f}\left[1 + \frac{K_a}{(A)}\left(1 + \frac{(Q)}{K_{iq}}\right)\right] \tag{6.28}$$

At a subsaturating concentration of the non-varied substrate, Q is a mixed type inhibitor when B is the variable substrate. If the enzyme is saturated with A, Q exerts no inhibitory effect. The product inhibition pattern of the ping-pong reaction sequence is presented in Figure 6.6. Product inhibition is not required to distinguish the ping-pong reaction sequence from the other two models in Figure 5.1, but there are reaction sequences which exhibit the same kinetic behavior as the ping-pong in the absence of products. One of these will be considered in the following chapter.

6.3 A reaction sequence with abortive complexes

The reaction sequences which have been considered up to this point have consisted of a catalytic cycle without the formation of abortive complexes.

This is a common phenomenon among pyridine nucleotide-linked dehydrogenases. In the case of these enzymes, the oxidized or reduced nucleotide is usually the first substrate to bind to, and the last product to dissociate from, the enzyme. To illustrate, after NADH has bound to the dehydrogenase, the oxidized second substrate binds in the normal reaction sequence. This is followed by the transfer of electrons to the second substrate followed by dissociation of the reduced second substrate and then dissociation of the NAD$^+$ from the dehydrogenase. However, if the reduced second substrate (product) is present in the assay medium, it may bind to the E-NADH complex to form an unproductive (abortive) E-NADH-reduced substrate complex. In like manner, the oxidized second substrate may bind to the E-NAD$^+$ complex to form an unproductive E-NAD$^+$-oxidized substrate complex.

The reaction sequence for this enzyme with two abortive complexes is portrayed in Figure 6.7. The connection matrix and the Q matrix for this reaction sequence are presented in Figure 6.8. The enzyme distribution expressions for this reaction sequence are given in Figure 6.9. The reader should note that there are a total of six elements in the vector for each term in the distribution expressions. This reflects the fact that there are six enzymes species in the reaction sequence. However, there are a total of only four terms in the distribution expression for each enzyme species, and this is a reflection of the fact that there are only four enzyme species in the actual catalytic cycle. The enzyme species which constitute the catalytic cycle are E, EA, EAB, and EQ. Equation (6.29) is the complete rate equation in the coefficient vector form for the reaction sequence presented in Figure 6.7.

The enzyme distribution expressions for the reaction sequence are given in Fig. 6.9. Equation (6.29) is the complete rate equation in the coefficient form.

$$
v = \frac{[(2,3,4,1,2,4)(A)(B) - (4,1,2,3,2,4)(P)(Q)}{\begin{array}{l} (0,1,2,1,2,4) \quad (2,3,4,0,2,4)\,(A)\,(B) \\ (0,1,4,1,2,4) \quad (2,3,0,1,2,4)\,(A)\,(B) \;\; (2,3,4,6,2,0)\,(A)\,(B)^2 \\ (2,0,2,1,2,4)\,(A) \quad (2,5,2,1,0,4)\,(A)\,(P) \;\; (2,3,0,3,2,4)\,(A)\,(B)\,(P)^2 \\ (2,0,4,1,2,4)\,(A) \quad (2,5,4,1,0,4)\,(A)\,(P) \\ \qquad\qquad\qquad\quad\; (2,0,2,3,2,4)\,(A)\,(P) \;\; (2,5,2,3,0,4)\,(A)\,(P)^2 \\ (0,3,4,1,2,4)\,(B) \quad (4,1,2,6,2,0)\,(B)\,(Q) \;\; (4,3,4,6,2,0)\,(B)^2(Q) \\ (0,1,2,3,2,4)\,(P) \quad (4,1,4,6,2,0)\,(B)\,(Q) \\ \qquad\qquad\qquad\quad\; (4,3,4,0,2,4)\,(B)\,(Q) \;\; (4,3,0,3,2,4)\,(B)\,(P)\,(Q) \\ (4,1,2,0,2,4)\,(Q) \quad (4,1,0,3,2,4)\,(P)\,(Q) \;\; (4,5,2,3,0,4)\,(P)^2(Q) \\ (4,1,4,0,2,4)\,(Q) \quad (4,1,2,3,2,4)\,(P)\,(Q) \end{array}}
$$

(6.29)

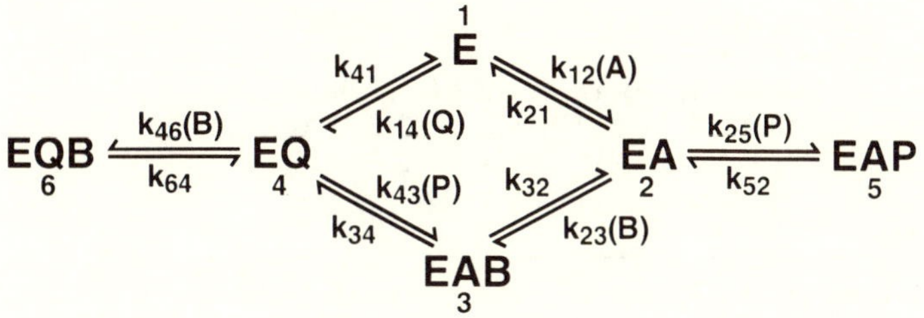

Fig. 6.7. Reaction sequence of an ordered, bi-bi, sequential model with two abortive complexes.

$$
U = \begin{vmatrix}
0 & A & 0 & Q & 0 & 0 \\
1 & 0 & B & 0 & P & 0 \\
0 & 1 & 0 & 1 & 0 & 0 \\
1 & 0 & P & 0 & 0 & B \\
0 & 1 & 0 & 0 & 0 & 0 \\
0 & 0 & 0 & 0 & 0 & 0
\end{vmatrix}
\qquad
Q = \begin{vmatrix}
2 & 4 & 0 \\
1 & 3 & 5 \\
2 & 4 & 0 \\
1 & 3 & 6 \\
2 & 0 & 0 \\
4 & 0 & 0
\end{vmatrix}
$$

Fig. 6.8. Connection matrix and matrix of non-zero elements for the model in Figure 6.7.

Reformulation of eq. (6.29) into the kinetic form gives,

$$
v = \cfrac{V_f\left[1 - \cfrac{\Gamma}{K_{eq}}\right]}{\begin{aligned}&1 + \frac{K_a}{(A)} + \frac{K_b}{(B)} + \frac{K_{ia_1}K_b}{(A)(B)} + \frac{(B)}{K_{ib_1}} + \frac{(P)}{K_{ip_2}} + \frac{K_a(P)}{K_{ip_1}(B)} + \frac{K_{ia_2}K_b(P)}{K_{ip_1}(A)(B)} \\[2mm] &+ \frac{K_a(P)^2}{K_{ip_1}K_{ip_3}(B)} + \frac{K_a(Q)}{K_{iq_2}(A)} + \frac{K_{ia_2}K_b(Q)}{K_{iq_1}(A)(B)} + \frac{K_a(B)(Q)}{K_{ib_2}K_{iq_2}(A)} \\[2mm] &+ \frac{K_a(P)(Q)}{K_{ip_4}K_{iq_2}(A)} + \frac{K_{ia_1}K_b(P)(Q)}{K_p K_{iq_1}(A)(B)} + \frac{K_{ia_1}K_b(P)^2(Q)}{K_p K_{ip_5}K_{iq_1}(A)(B)}\end{aligned}}
$$

(6.30)

A couple of features become apparent upon inspection of eq. (6.30). Firstly, substrate B will give rise to substrate inhibition. This is due to the fifth denominator term and also because of the twelfth denominator term, although, in the case of the latter term, inhibition requires the presence of product Q in the assay medium. This is distinct from the explanation of substrate inhibition discussed in Chapter 3. Secondly, product inhibition

(E)/E	(EA)/E	(EAB)/E
$(0,1,2,1,2,4)$	$(2,0,2,1,2,4)(A)$	$(2,3,0,1,2,4)(A)(B)$
$(0,1,2,3,2,4)(P)$	$(2,0,2,3,2,4)(A)(P)$	$(2,3,0,3,2,4)(A)(B)(P)$
$(0,1,4,1,2,4)$	$(2,0,4,1,2,4)(A)$	$(4,1,0,3,2,4)(P)(Q)$

$(EQ)/E_t$	$(EAP)/E_t$	$(EQB)/E_t$
$(2,3,4,0,2,4)(A)(B)$	$(2,5,2,1,0,4)(A)(P)$	$(2,3,4,6,2,0)(A)(B)^2$
$(4,1,2,0,2,4)(Q)$	$(2,5,2,3,0,4)(A)(P)^2$	$(4,1,2,6,2,0)(B)(Q)$
$(4,1,4,0,2,4)(Q)$	$(2,5,4,1,0,4)(A)(P)$	$(4,1,4,6,2,0)(B)(Q)$
$(4,3,4,0,2,4)(B)(Q)$	$(4,5,2,3,0,4)(P)^2(Q)$	$(4,3,4,6,2,0)(B)^2(Q)$

Fig. 6.9. Enzymes distribution expressions for the enzyme reaction sequence in Figure 6.7.

by product P is a second order function of the concentration of P. The Lineweaver-Burk equations in the absence of the products are the following.

$$\frac{1}{v_a} = \frac{K_a}{V_f}\left[1 + \frac{K_{ia_1}K_b}{K_a(B)}\right]\frac{1}{(A)} + \frac{1}{V_f}\left[1 + \frac{K_b}{(B)} + \frac{(B)}{K_{ib_1}}\right] \tag{6.31}$$

$$\frac{1}{v_b} = \frac{K_b}{V_f}\left[1 + \frac{K_{ia_1}}{(A)}\right]\frac{1}{(B)} + \frac{1}{V_f}\left[1 + \frac{K_a}{(A)} + \frac{(B)}{K_{ib_1}}\right] \tag{6.32}$$

Substrate inhibition by the second substrate is manifested by an increase in the intercept of the Lineweaver-Burk plot regardless of which substrate is the variable substrate. When P is the product inhibitor, the following Lineweaver-Burk equations are obtained.

$$\frac{1}{v_a} = \frac{K_a}{V_f}\left[1 + \frac{K_{ia_1}K_b}{K_a(B)}\left(1 + \frac{K_{ia_2}(P)}{K_{ia_1}K_{ip_1}}\right)\right]\frac{1}{(A)}$$
$$+ \frac{1}{V_f}\left[1 + \frac{K_b}{(B)}\left(1 + \frac{(P)}{K_{ip_1}} + \frac{(P)^2}{K_{ip_1}K_{ip_3}}\right) + \frac{(P)}{K_{ip_2}} + \frac{(B)}{K_{ib_1}}\right] \tag{6.33}$$

$$\frac{1}{v_b} = \frac{K_b}{V_f}\left[1 + \frac{K_{ia_1}}{(A)}\left(1 + \frac{K_{ia_2}(P)}{K_{ia_1}K_{ip_1}}\right) + \frac{(P)}{K_{ip_1}} + \frac{(P)^2}{K_{ip_1}K_{ip_3}}\right]\frac{1}{(B)}$$
$$+ \frac{1}{V_f}\left[1 + \frac{K_a}{(A)} + \frac{(P)}{K_{ip_2}} + \frac{(B)}{K_{ib_1}}\right] \tag{6.34}$$

Inspection of eqs. (6.33) and (6.34) provides some valuable information about the location of the abortive complex EAP in the reaction sequence. The term which contains $(P)^2$ is associated with the EAP complex. This term affects the intercept of eq. (6.33) where A is the variable substrate.

Thus, the EAP complex is analogous to the EAI complex in uncompetitive inhibition. Therefore, the reaction which gives rise to the EAP complex occurs after the binding of substrate A to the enzyme. The secondary plot of the intercepts versus (P) when A is the variable substrate and B is present at a sub-saturating concentration will not be linear, but rather, parabolic. With respect to saturation with B, it should be noted that for the enzyme sequence under consideration it is not possible to saturate the enzyme with B. Equation (6.34) shows that, when B is the variable substrate, the term $(P)^2$ term is in the slope component of the Lineweaver-Burk equation. Thus, the reaction which gives rise to the EAP complex is competitive with the interaction of B with the enzyme during the normal reaction sequence. The secondary plot of the slopes of the Lineweaver-Burk plots as a function of the concentration of P will be parabolic regardless of the concentration of A. The non-linearity of the secondary plots are evidence that abortive complexes are formed during the reaction sequence.

6.4 Problems for Chapter 6

6.1 Utilizing the equation derived in problem 5.1 of chapter 5, write the Lineweaver-Burk equations for each of the substrates with respect to each of the products.

6.2 Prepare a product inhibition pattern for the enzyme model presented in problem 5.1 of chapter 5.

6.3 Utilizing the equation derived in problem 5.2 of chapter 5, write the Lineweaver-Burk equations for each of the substrates with respect to each of the products.

6.4 Prepare a product inhibition pattern for the enzyme model portrayed in problem 5.2 of chapter 5.

References

1. Cleland, W. W. (1963). The kinetics of enzyme-catalyzed reactions with two of more substrates or products. II. Inhibition: Nomenclature and theory. *Biochim. Biopys. Acta* **67**: 173–87.
2. Cleland, W. W. (1967). Enzyme kinetics. *Ann. Rev. Biochem.* **36**: 77–112.
3. Segel, I. H. (1975). *Enzyme Kinetics*, pp. 506–845, New York, John Wiley & Sons.
4. Dixon, M. and Webb, E. C. (1979). *Enzymes*, 3rd. Ed., pp. 47–137, New York, Academic Press.

7

Prediction of reaction sequence

The methods discussed in previous chapters provide for the derivation of rate equations of even complex enzyme-catalyzed reactions. It is true that, if the reaction sequence of the enzymic reaction is highly random, a computer-based derivation is the only feasible alternative, but the algorithm described in chapter 5 can be incorporated into a computer-based procedure. However, there is a problem with the approach that has been followed to this point. The underlying presumption for each enzymic reaction has been that there is a *known* reaction sequence. If the reaction sequence is known, regardless of how complex that sequence may be, it is possible to derive an appropriate rate equation. Herein lies the problem! When one initiates an investigation of the kinetic behavior of an enzyme, the reaction sequence is not known, and the presumption of a plausible reaction sequence is the *worst* of the possible approaches to the investigation.

7.1 The enzyme kineticist and the mystery novel

The unravelling of the reaction sequence of an enzymic reaction is completely analogous to the unravelling of a crime in a mystery novel. The study of the kinetic behavior of an enzyme has all the challenge and intrigue of a well-written mystery novel! While instinct may be of some small value to the detective who would determine the kinetic behavior of an enzyme or solve a murder mystery, instinct can prove to be a fickle ally. Precise investigative work requires the combination of asking the proper questions at the proper time and a sagacious interpretation of the answers received. The enzyme kineticist has only one "witness" to whom the questions can be addressed, and that is the enzyme itself. This both simplifies and complicates the problem. It is simplified because it "narrows the field," but it complicates the problem because it lessens, but does not eliminate, the possibility of obtaining corroborating evidence.

The questions which the enzyme kineticist must address to the enzyme relate to the kinetic behavior of the enzyme. More often than not, this consists of conducting substrate-saturation experiments under carefully defined conditions. The enzyme provides the answers in the form of the substrate-saturation curves. These curves must be analyzed and interpreted meticulously. In chapter 4 and again in chapter 6, it was noted that, under certain conditions, a product might serve as a competitive inhibitor of a given substrate. In each case, it was implied that this was an expected result because "the substrate and product interact with the same species of the enzyme." This means that, if it can be shown that two reactants interact with the enzyme in a competitive manner, it can be concluded that the reaction sequence is such that it provides for those two reactants to react with the same species of the enzyme in a mutually exclusive manner. The discussion of the ordered, sequential reaction sequence with the formation of abortive complexes in the latter part of the previous chapter provides additional examples of interpretations that can be made on the basis of studies of the kinetic behavior of the enzyme.

Cleland[1] has proposed rules which are extremely valuable in the interpretation of enzyme kinetic data. These rules will be stated here as follows.

1) If a reactant affects the intercept of the Lineweaver-Burk plot of the variable substrate, that reactant reacts with a different species of the enzyme than does the variable substrate. Thus, if the reactant reacts with a species other than that species with which the variable substrate reacts, the reactant will affect the maximum velocity.

2) If a reactant affects the slope of the Lineweaver-Burk plot, that reactant either reacts with the same species of the enzyme as does the variable substrate *or* it reacts with a species of the enzyme which is connected to the species of the enzyme with which the variable substrate reacts by a *reversible* path. If the latter criterion is met, the reactant will affect V_f^{app}/K_a^{app}, the efficiency of the enzyme. With regard to the second rule, it is important to recognize that the term "reversible" does *not* imply equilibrium. Furthermore, it must be understood that there are three conditions which could result in the irreversibility of an individual step in the reaction sequence. The most obvious is that a given step might be essentially irreversible because of thermodynamic considerations. Secondly, a step which involves the interaction of the enzyme with a reactant would be irreversible if the concentration of that reactant were equal to zero. Thus, in the absence of a product, the step in which that product reacts with the

enzyme is irreversible. Finally, saturation of the enzyme with a substrate renders the step at which that substrate reacts with the enzyme irreversible due to mass action considerations. The latter two conditions constitute kinetic irreversibility rather than thermodynamic irreversibility.

7.2 Product inhibition patterns: The clues with which the enzyme kineticist works

In the light of the foregoing considerations, the product inhibition pattern of the ordered, sequential enzyme model can be considered, and it will be seen that if an enzyme of an unknown reaction sequence were to exhibit that kinetic behavior it would be possible to state that the kinetic behavior of the enzyme is consistent with that of an ordered, sequential sequence. Before the product inhibition pattern of Figure 7.1 is considered, it is advantageous to remember that, in the absence of either product, the non-varied substrate affected both the slope and intercept of the Lineweaver-Burk plots, in the case of the ordered, sequential reaction sequence. The fact that the intercept was affected showed that the non-varied substrate interacted with a different species of the enzyme than did the variable substrate. The fact that the slope was affected indicated that the species of the enzyme with which the two substrates interact were connected by a reversible path. The simplest interpretation of this information is either of the sequences shown in Figure 7.2. The first row in the product inhibition pattern portrayed in Figure 7.1 shows that, when P was the product inhibitor and the non-varied substrate was present at a subsaturating concentration, the product affected both the slope and the intercept of the Lineweaver-Burk plot. Since the intercept is affected, P interacts with a different species of the enzyme

(A)	(B)	Inhibitor	Type of inhibition
Variable	Subsaturate	P	Mixed type
Variable	Saturate	P	Uncompetitive
Subsaturate	Variable	P	Mixed type
Saturate	Variable	P	Mixed type
Variable	Subsaturate	Q	Competitive
Variable	Saturate	Q	Competitive
Subsaturate	Variable	Q	Mixed type
Saturate	Variable	Q	No inhibition

Fig. 7.1. Product inhibition pattern for the enzyme reaction sequence in Figure 5.1A.

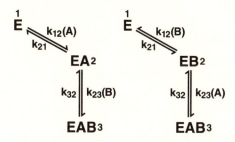

Fig. 7.2. Two possible partial reaction sequences based on observations of a sequential reaction sequence in the absence of products.

than does substrate A. The fact that the slope was affected indicated that, at a subsaturating concentration of the non-varied substrate, the path between the species with which A and P react was reversible. Saturation of the enzyme with B changes the type of inhibition by P such that the slopes of the Lineweaver-Burk plots are not affected. This suggests that B interacts with the enzyme *after* the variable substrate since a reversible path between the interaction of A and P with the enzyme is no longer there. When B was the variable substrate, the product P was a mixed type inhibitor regardless of whether or not the enzyme was saturated with A. Thus, P reacted with a different species of the enzyme than did B, and the path between the species with which B and P react was reversible regardless of the concentration of A. These results are consistent with the sequence shown in Figure 7.3. The fourth enzyme species is called EX because, at this point in the analysis, there is not sufficient information available to determine the nature of this species. However, when A was the variable substrate, the product Q was a competitive inhibitor. That is, Q affected the slope of the Lineweaver-Burk plots but it had no affect on the intercepts. Thus, substra-

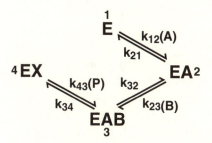

Fig. 7.3. Reaction sequence consistent with the product inhibition pattern in Figure 7.1 when the product inhibitor is P and Q is absent.

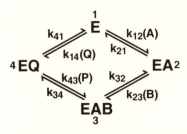

Fig. 7.4. Reaction sequence consistent with the product inhibition pattern in Figure 7.1.

te A and product Q interacted with the same species of the enzyme. When B was the variable substrate and A was present at a sub-saturating concentration, Q affected both the slopes and the intercepts of the Lineweaver-Burk plots. Hence, substrate B and product Q interacted with different species of the enzyme but these species were connected by a reversible path even when product P was absent. However, when the enzyme was saturated with A, there was no enzyme with which Q could interact. The concentration of A required to saturate the enzyme in the presence of Q would be greater than in the absence of Q because, as shown in Chapter 3, a competitive inhibitor increases the apparent Michaelis constant of the variable substrate, but if the enzyme is saturated with A there is no enzyme with which Q can interact. Figure 7.4 shows the simplest reaction sequence which is consistent with the observed results. The choice of the words, "consistent with the observed results," is significant. The results obtained in studies of the kinetic behavior of a system are not unique, that is, they can usually be interpreted in more than one way. Actually, this is true of many facets of science. It is very difficult to *prove* beyond question that any hypothesis is correct. In most cases, a more precise claim is that all the available evidence is consistent with a given hypothesis. This is true of an astute detective in a mystery novel as well as for the astute enzyme kineticist!

Cleland's rules can be utilized to predict a reaction sequence on the basis of the product inhibition pattern of the ordered, ping-pong model presented in Figure 5IB. Before the product inhibition pattern shown in Figure 7.5 is considered, it should be recalled that, in the case of the ping-pong model, in the absence of either product, the non-varied substrate affected the intercept but not the slope of the Lineweaver-Burk plots. In accordance with Cleland's rules, this indicates that the non-varied substrate reacted with a different species of the enzyme than did the variable substrate and that

(A)	(B)	Inhibitor	Type of inhibition
Variable	Subsaturate	P	Mixed type
Variable	Variable	P	No inhibition
Subsaturate	Variable	P	Competitive
Saturate	Variable	P	Competitive
Variable	Subsaturate	Q	Competitive
Variable	Saturate	Q	Competitive
Subsaturate	Variable	Q	Mixed type
Saturate	Variable	Q	No inhibition

Fig. 7.5. Product inhibition pattern for the enzyme reaction sequence in Figure 5.1B.

there was not a reversible path between these two enzyme species in the absence of products. When A was the variable substrate and B was present at a subsaturating concentration, P affected both the slope and the intercept of the Lineweaver-Burk plots. This indicates that A and P reacted with different species of the enzyme and that there was a reversible path between these species, under the conditions specified. Because saturation of the enzyme with the non-varied substrate eliminated inhibition by P, one would predict that substrate B and product P reacted with the same species of the enzyme, and this is confirmed by the fact that P inhibited the enzyme competitively when B was the variable substrate. These observations suggest the following reaction sequence. The product Q was a competitive inhibitor when A was the variable substrate, and therefore, these reactants interacted with the same species of the enzyme. When B was the variable substrate, Q was a mixed type inhibitor if A was present at a subsaturating concentration, and this indicates that, under these conditions, the path between the enzyme species with which B and Q react was reversible.

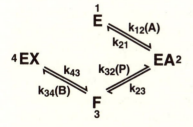

Fig. 7.6. Reaction sequence consistent with the product inhibition pattern in Figure 7.5 when P is the product inhibitor and Q is absent.

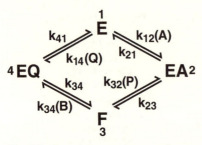

Fig. 7.7. Reaction sequence consistent with the entire product inhibition pattern in Figure 7.5.

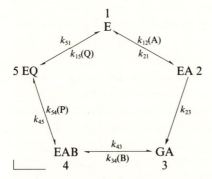

Fig. 7.8. Model of an ordered, bi-bi, sequential reaction sequence in which an irreversible step intervenes between the addition of the two substrates. EA and GA are different forms of the enzyme-substrate complex.

Saturation with A, of course, eliminated inhibition by Q. These findings support the reaction sequence presented in Figure 7.7.

The observations with the ping-pong model illustrate an important point. In the absence of products, there is no reversible path between the enzyme species with which substrates A and B react. However, in the presence of a subsaturating concentration of either product, the path between the species with which the two substrates interact is reversible. This means that if a series of substrate-saturation experiments were conducted each in the presence of a subsaturating concentration of either product, the non-varied substrate would affect both the slope and the intercept of the Lineweaver-Burk plots. Furthermore, there are sequential models which will mimic the behavior of a ping-pong model in the absence of products. Consider the following sequential reaction in Figure 7.8. The reaction sequence is irreversible. This might be due to the fact that the EA species is

(A)	(B)	Inhibitor	Type of inhibition
Variable	Subsaturate	P	Uncompetitive
Variable	Saturate	P	Uncompetitive
Subsaturate	Variable	P	Mixed type
Saturate	Variable	P	Mixed type
Variable	Subsaturate	Q	Competitive
Variable	Saturate	Q	Competitive
Subsaturate	Variable	Q	Uncompetitive
Saturate	Variable	Q	No inhibition

Fig. 7.9. Product inhibition pattern for the reaction sequence in Figure 7.8

unstable in the assay medium and is irreversibly converted to GA. If a series of substrate-saturation experiments were conducted at different concentrations of the non-varied substrate in the absence of products, it would be found that the non-varied substrate would affect the intercept but not the slope of the Lineweaver-Burk plots. On the basis of this observation, one might conclude erroneously that the reaction sequence is a ping-pong mechanism in which the first product dissociates from the enzyme before the second substrate binds. However the predicted product inhibition pattern illustrated in Figure 7.9 is distinct from the pattern shown in Figure 7.9 for the ping-pong model. The foregoing illustrates both the danger of drawing conclusions on the basis of insufficient evidence and the power of product inhibition to distinguish between reaction sequences.

The ordered, sequential reaction sequence and the ordered, iso-sequential sequence in the previous chapter were found to differ only with respect to the product inhibition pattern with respect to product Q. The product inhibition pattern for the iso-sequential model when Q is the product inhibitor is shown in Figure 7.10. When A was the variable substrate, Q was a mixed type inhibitor regardless of the concentration of the non-varied substrate. This is a reflection of the fact that A and Q reacted with different species of the enzyme and that the path between these species in the model

(A)	(B)	Inhibitor	Type of inhibition
Variable	Subsaturate	Q	Mixed type
Variable	Saturate	Q	Mixed type
Subsaturate	Variable	Q	Mixed type
Saturate	Variable	Q	Uncompetitive

Fig. 7.10. Product inhibition pattern for the reaction sequence in Figure 5.1C when Q is the product inhibitor and P is absent.

was reversible. It is clear from Figure 7.10 that substrate B reacted with a different species of the enzyme than product Q, and that the path between these species was reversible if A was present at a subsaturating concentration, but irreversible if the enzyme were saturated with A. This indicates that A interacts with the enzyme at a point in the reaction sequence between the dissociation of Q from the enzyme and the binding of B. When these observations are combined with the appropriate evidence summarized in Figure 7.5, the reaction sequence portrayed in Figure 5.1C can be predicted.

It was pointed out at the beginning of this chapter that the task facing the enzyme kineticist is analogous to that of a detective who would solve a crime. In both cases, it is important that the investigator initiate the investigation free from the bias of a preconceived hypothesis. In both cases the investigator must ask incisive questions and must evaluate the evidence obtained by this procedure in a judicious and meticulous manner. In the case of enzyme kinetics, this requires that at each step in the procedure the investigator consider all of the reaction models that his imagination can envision. The rate equation for all of these reaction models should be derived, because the terms present in each of these equations will provide the kineticist with the questions which must be addressed to the enzyme. Discrimination between two reaction sequences may depend on the utilization of the information available in only one of numerous denominator terms. Finally, the enzyme kineticist must always be cognizant of the fact that there are no absolute conclusions; one can only demonstrate that the observed results are consistent with a given reaction sequence. Thus the kineticist obtains only circumstantial evidence. It should also be realized that Cleland's rules apply to the catalytic reaction sequence and not to abortive complexes. However, the existence of abortive complexes can usually be detected by nonlinear secondary plots, and evidence concerning the location of an abortive complex can often be discerned by careful analysis[2]. Segel[3] has discussed many additional reaction sequences.

7.3 Problems for chapter 7

7.1 The following product inhibition pattern was obtained in a series of substrate-saturation experiments involving an enzyme which catalyzes the following reaction

$$A + B \rightleftharpoons P + Q + R.$$

(A)	(B)	Inhibitor	Type of inhibition
Variable	Subsaturated	P	Uncompetitive
Variable	Saturated	P	No inhibition
Subsaturated	Variable	P	Competitive
Variable	Subsaturated	Q	Mixed type
Variable	Saturated	Q	Mixed type
Subsaturated	Variable	Q	Uncompetitive
Variable	Subsaturated	R	Competitive
Subsaturated	Variable	R	Mixed type
Saturated	Variable	R	No inhibition

Propose a reaction sequence which is consistent with the foregoing observations.

7.2 An enzyme which catalyzes the following reaction

$$A + B + C \rightleftharpoons P + Q$$

gave rise to the following product inhibition pattern.

(A)	(B)	(C)	Inhibitor	Type of inhibition
Variable	Subsat.	Subsat.	P	Mixed type
Variable	Sat.	Subsat.	P	Mixed type
Variable	Subsat.	Sat.	P	Mixed type
Subsat	Variable	Subsat.	P	Mixed type
Sat.	Variable	Subsat.	P	Uncompetitive
Subsat.	Variable	Sat.	P	Mixed type
Subsat.	Subsat.	Variable	P	Mixed type
Sat.	Subsat.	Variable	P	Uncompetitive
Subsat.	Sat.	Variable	P	Uncompetitive
Variable	Subsat.	Subsat.	Q	Mixed type
Variable	Sat.	Subsat.	Q	Uncompetitive
Variable	Subsat.	Sat.	Q	Uncompetitive
Subsat.	Variable	Subsat.	Q	Mixed type
Sat.	Variable	Subsat	Q	Mixed type
Subsat.	Variable	Sat.	Q	Uncompetitive
Subsat.	Subsat.	Variable	Q	Mixed type
Sat.	Subsat.	Variable	Q	Mixed type
Subsat.	Sat.	Variable	Q	Mixed type

Propose a reaction sequence which is consistent with the foregoing observations.

References

1. Cleland, W. W. (1963). The kinetics of enzyme-catalyzed reactions with two or more substrates or products. III. Prediction of initial velocity and inhibition patterns by inspection. *Biochim. Biophys. Acta* **67**: 188–96.
2. Davisson, V. J. and Schulz, A. R. (1985). The purification and steady-state kinetic behaviour of rabbit heart mitochondrial NAD(P)$^+$ malic enzyme. *Biochem. J.* **225**: 335–42.
3. Segel, I. H. (1975). *Enzyme Kinetics*, pp. 505–845, New York, John Wiley & Sons.

8

Enzyme-catalyzed isotopic exchange

Isotopes have been used extensively in enzymology to obtain answers to a number of different types of questions. Thus, isotopes have been used to gain insight into the mechanism of the catalytic reaction[1]. Isotopes have also been used to locate the amino acids which constitute the active site[2] or the allosteric site[3] of enzymes. The following chapter will contain a discussion of how the isotope effect can be utilized to obtain information concerning the location of rate limiting steps in an enzyme sequence. The present chapter will be concerned with the study of the kinetics of isotopic exchange as a means of 'fine tuning' the understanding of the reaction sequence of an enzyme. It will be assumed in the discussion in this chapter that there is no significant isotope effect. That is, that the mass of the label isotope does not differ from the normal isotope significantly. Furthermore, it will be assumed that the concentration of the isotope is very small compared to the concentration of the normal isotope.

Isotopic exchange provides a means of obtaining an insight into the reaction sequence of an enzymic reaction which, in some instances, would be difficult or impossible to obtain by the methods described in the previous chapters. While the information available from analysis of isotopic exchange data can be very useful, the belief that these data are directly related to observations obtained in initial rate studies can be erroneous. This fact was pointed out by Boyer[1] in his classic paper on isotopic exchange under equilibrium conditions. Most isotopic exchange experiments are conducted under equilibrium conditions, and most of the formal treatments have dealt with the equilibrium[1-3]. However, some treatments have dealt with steady state isotopic exchange[4-6].

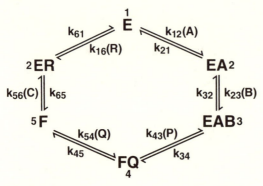

Fig. 8.1. Reaction sequence of an ordered, bi-bi, sequential, uni-uni, ping-pong enzyme reaction.

$(E)/E_t$	$(EA)/E_t$	$(EAB)/E_t$
$(0, 1, 2, 3, 4, 1)(P)(Q)$	$(2, 0, 2, 3, 4, 1)(A)(P)(Q)$	$(2, 3, 0, 3, 4, 1)(A)(B)(P)(Q)$
$(0, 1, 2, 3, 4, 5)(P)(Q)$	$(2, 0, 2, 3, 4, 5)(A)(P)(Q)$	$(2, 3, 0, 3, 4, 5)(A)(B)(P)(Q)$
$(0, 1, 2, 3, 6, 1)(C)(P)$	$(2, 0, 2, 3, 6, 1)(A)(C)(Q)$	$(2, 3, 0, 3, 6, 1)(A)(B)(C)(P)$
$(0, 1, 2, 5, 6, 1)(C)(Q)$	$(2, 0, 2, 5, 6, 1)(A)(C)$	$(2, 3, 0, 5, 6, 1)(A)(B)(C)$
$(0, 1, 4, 5, 6, 1)(C)$	$(2, 0, 4, 5, 6, 1)(A)(C)$	$(6, 1, 0, 3, 4, 5)(P)(Q)(R)$
$(0, 3, 4, 5, 6, 1)(B)(C)$	$(6, 0, 2, 3, 4, 5)(P)(Q)(R)$	$(6, 3, 0, 3, 4, 5)(B)(P)(Q)(R)$

$(FQ)/E_t$	$(F)/E_t$	$(ER)/E_t$
$(2, 3, 4, 0, 4, 1)(A)(B)(Q)$	$(2, 3, 4, 5, 0, 1)(A)(B)$	$(2, 3, 4, 5, 6, 0)(A)(B)(C)$
$(2, 3, 4, 0, 4, 5)(A)(B)(Q)$	$(2, 3, 4, 5, 0, 5)(A)(B)$	$(6, 1, 2, 3, 4, 0)(P)(Q)(R)$
$(2, 3, 4, 0, 6, 1)(A)(B)(C)$	$(6, 1, 2, 3, 0, 5)(P)(R)$	$(6, 1, 2, 3, 6, 0)(C)(P)(R)$
$(6, 1, 2, 0, 4, 5)(Q)(R)$	$(6, 1, 2, 5, 0, 5)(R)$	$(6, 1, 2, 5, 6, 0)(C)(R)$
$(6, 1, 4, 0, 4, 5)(Q)(R)$	$(6, 1, 4, 5, 0, 5)(R)$	$(6, 1, 4, 5, 6, 0)(C)(R)$
$(6, 3, 4, 0, 4, 5)(B)(Q)(R)$	$(6, 3, 4, 5, 0, 5)(B)(R)$	$(6, 3, 4, 5, 6, 0)(B)(C)(R)$

Fig. 8.2. Enzyme distribution expressions for the enzyme reaction in Figure 8.1.

8.1 Isotopic exchange in an ordered reaction sequence

The enzymic reaction sequence which will be employed for this discussion is the sequence considered by Cleland[5]. The following are the distribution expressions for this enzyme-catalyzed reaction.

The treatment of isotopic exchange which will be presented in this text is the steady state treatment[5,6]. This is, in fact, a bit misleading because it is necessary to measure the rate of isotopic exchange under conditions where the rate of exchange is measured in the absence of a net reaction. Since the reaction sequence portrayed in Figure 8.1 includes a ping-pong sequence, it is possible to measure an exchange between labelled A and unlabelled Q, for example, in the absence of a net reaction by simply excluding substrate

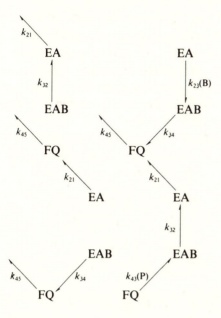

$$N_t = k_{21}k_{45}[k_{32} + k_{34}] + k_{23}k_{34}k_{45}(B) + k_{21}k_{32}k_{43}(P)$$

Fig. 8.3. Paths involved in A → Q and Q → A exchange for the reaction sequence in Fig. 8.1. Only those enzyme species which become labelled with isotope are shown.

C and product R from the reaction medium. However, in this case, there will be an equilibrium established between the enzyme species involved prior to the binding of C to the enzyme. The steady state treatment will be presented here because it follows a progression which is more consistent with the treatment given previously in this text. The derivation of an expression for the rate of isotopic exchange can be illustrated by considering the rate of exchange between $\dot{A}$ and Q where $\dot{A}$ indicates that A is labelled initially. The rate of isotopic exchange is given by eq. (8.1).

$$\dot{v}_{a \to q} = \left[k_{12}(\dot{A}) \frac{N_{fq}}{N_t} \frac{(E)}{E_t} \right] E_t \qquad (8.1)$$

where N_t is the total number of paths which can be followed by isotopically labelled enzyme species and N_{fq} is the path which connects all the labelled species of the enzyme directly with FQ, which is the labelled species which dissociates to form labelled Q. The paths which lead to the exchange between the species of the enzyme which become labelled are shown in Figure 8.3.

If the rate of exchange between A and Q is to be investigated, it is certain that both A and Q must be present in the initial reaction medium. Equation 8.1 contains the concentration of A, and N_{fq} contains the concentration of B since $N_{fq} = k_{23}k_{34}k_{45}$(B). The distribution expression for (E)/E_t, as shown in Figure 8.2 contains two terms which contain (Q), but they also contain (P). The remaining terms in (E)/E_t contain (C), and the concentration of C must be set to zero to prevent a net synthesis of Q. Thus, when the appropriate terms for N_{fq}/N_t and (E)/E_t are substituted into eq. (8.1), the numerator will contain the concentrations of A, B, P, and Q, but C and R are not required. A substitution of these terms into eq. (8.1) gives

$$\dot{v}_{a \to q} = \frac{k_{12}(k_{23}k_{34}k_{45})(k_{21}k_{32}k_{43}k_{54}(k_{61}+k_{65}))(A)(B)(P)(Q)E_t}{[k_{21}k_{45}(k_{32}+k_{34})+k_{23}k_{34}k_{45}(B)+k_{21}k_{32}k_{43}(P)]\dot{D}} \tag{8.2}$$

The symbol $\dot{D}$ in eq. (8.2) is the sum of the distribution expressions shown in Figure 8.3 *excluding* those terms which contain either (C) or (R). Therefore, $\dot{D}$ can be represented as

$$\dot{D} = \text{coef. AB}\,(A)(B) + \text{coef. PQ}\,(P)(Q) + \text{coef. ABQ}\,(A)(B)(Q)$$
$$+ \text{coef. APQ}\,(A)(P)(Q) + \text{coef. ABPQ}\,(A)(B)(P)(Q) \tag{8.3}$$

Furthermore, the three terms in the denominator of eq. (8.2) which constitute N_t can be expressed as constant + coef. B(B) + coef. P(P). Multiplication of the numerator and denominator of eq. (8.2) by the factor $k_{56}k_{61}/$ coef. ABC allows for the following expression for the rate of isotopic exchange.

$$v_{a \to q} = \frac{V_f(A)(B)(P)(Q)E_t}{\dfrac{\text{constant} + \text{coef. B(B)} + \text{coef. P(P)}}{\text{coef. PQ} \times \text{coef. ABC}}\,\dot{D}} \tag{8.4}$$

The relevant steady state kinetic parameters are the following.

$$V_f = \frac{\text{num. 1}}{\text{coef. ABC}} = \frac{k_{34}k_{45}k_{61}E_t}{k_{34}k_{45}+k_{34}k_{61}+k_{45}k_{61}}$$

$$K_c = \frac{\text{coef. AB}}{\text{coef. ABC}} = \frac{k_{34}k_{45}(k_{61}+k_{65})}{k_{56}(k_{34}k_{45}+k_{34}k_{61}+k_{45}k_{61})}$$

$$K_{ia} = \frac{\text{coef. C}}{\text{coef. AC}} = \frac{\text{coef. CP}}{\text{coef. ACP}} = \frac{\text{coef. PQ}}{\text{coef. APQ}} = \frac{k_{21}}{k_{12}}$$

$$K_{ib_1} = \frac{\text{coef. APQ}}{\text{coef. ABPQ}} = \frac{\text{coef. ACP}}{\text{coef. ABCP}} = \frac{k_{32}}{k_{23}}$$

$$K_{ib_2} = \frac{\text{coef. C}}{\text{coef. BC}} = \frac{\text{coef. R}}{\text{coef. BR}} = \frac{\text{coef. CR}}{\text{coef. BCR}} = \frac{\text{coef. QR}}{\text{coef. BQR}} = \frac{k_{21}(k_{32} + k_{34})}{k_{23}k_{34}}$$

$$K_{ip_1} = \frac{\text{coef. ABQ}}{\text{coef. ABPQ}} = \frac{\text{coef. BQR}}{\text{coef. BPQR}} = \frac{k_{34}}{k_{43}}$$

$$K_{ip_2} = \frac{\text{coef. C}}{\text{coef. CP}} = \frac{\text{coef. R}}{\text{coef. PR}} = \frac{\text{coef. AC}}{\text{coef. ACP}} = \frac{\text{coef. CR}}{\text{coef. CPR}} = \frac{(k_{32} + k_{34})k_{45}}{k_{32}k_{43}}$$

$$K_{iq} = \frac{\text{coef. R}}{\text{coef. QR}} = \frac{\text{coef. AB}}{\text{coef. ABQ}} = \frac{\text{coef. BR}}{\text{coef. BQR}} = \frac{k_{45}}{k_{54}}$$

In addition to the foregoing steady state parameters, reformulation of eq. (8.4) into a kinetic form requires a definition of the following parameters which will be designated as exchange constants.

$$\dot{K}_a = \frac{\text{coef. B}}{\text{coef. AB}} = \frac{k_{56}k_{61}}{k_{12}(k_{61} + k_{65})}$$

$$\dot{K}_b = \frac{\text{constant}}{\text{coef. B}} = \frac{k_{21}(k_{32} + k_{34})}{k_{23} + k_{34}} = K_{ib_2}$$

$$\dot{K}_p = \frac{\text{constant}}{\text{coef. P}} = \frac{(k_{32} + k_{34})k_{45}}{k_{23} + k_{34}} = K_{ip_2}$$

$$\dot{K}_p = \frac{\text{coef. P}}{\text{coef. PQ}} = \frac{k_{56}k_{61}}{k_{54}(k_{61} + k_{65})}$$

Each term in the denominator of eq. (8.4) can be reformulated to the kinetic form by using the forgoing definitions. For example, the first term which involves a multiplication of coefficient AB by constant is reformulated in the following manner.

$$\frac{\text{constant} \times \text{coef. AB}}{\text{coef. ABC} \times \text{coef. PQ}} = \frac{\text{coef. AB}}{\text{coef. ABC}} \frac{\text{constant}}{\text{coef. P}} \frac{\text{coef. P}}{\text{coef. PQ}} = K_c K_{ip_2} \dot{K}_Q$$

The steady state equation for the rate of the $A \rightarrow Q$ exchange in the kinetic form is given in eq. (8.5).

$$V_f(\dot{A})(B)(P)(Q)/K_c\dot{D} \qquad (8.5)$$

where

$$\dot{D} = K_{ip_2}\dot{K}_q(\dot{A})(B) + \frac{K_{ip_2}\dot{K}_q}{K_{ib_2}}(\dot{A})(B)^2 + \dot{K}_q(\dot{A})(B)(P)$$

$$+ \dot{K}_a K_{ib_2}(P)(Q) + \dot{K}_a(B)(P)(Q) + \frac{\dot{K}_a K_{ib_2}}{K_{ip_2}}(P)^2(Q)$$

$$+ \frac{K_{ip_2}\dot{K}_q}{K_{iq}}(\dot{A})(B)(Q) + \frac{K_{ip_2}K_q}{K_{ib_2}K_{iq}}(\dot{A})(B)^2(Q) + \frac{\dot{K}_q}{K_{iq}}(\dot{A})(B)(P)(Q)$$

$$+ \frac{\dot{K}_a K_{ib_2}}{K_{ia}}(\dot{A})(P)(Q) + \frac{\dot{K}_a}{K_{ia}}(\dot{A})(B)(P)(Q) + \frac{\dot{K}_a K_{ib_2}}{K_{ia}K_{ip_2}}(\dot{A})(P)^2(Q)$$

$$+ \frac{K_{ip_2}\dot{K}_q}{K_{ip_1}K_{iq}}(\dot{A})(B)(P)(Q) + \frac{\dot{K}_a}{K_{ia}K_{ib_1}}(\dot{A})(B)^2(P)(Q)$$

$$+ \frac{\dot{K}_q}{K_{ip_1}K_{iq}}(\dot{A})(B)(P)^2(Q)$$

Equation (8.5) contains information which may not be obvious intuitively. The maximum velocity of the exchange is not equal to the maximum velocity of the overall reaction, but rather, it is equal to V_f/K_c. Furthermore, the rate of exchange is a 1:2 function of the concentrations of both B and P. While both B and P are required for the exchange to take place, a high concentration of either B or P will inhibit the exchange. If a sufficiently high concentration of either of these reactants is present, the exchange can be inhibited completely. The maximum rate of the exchange is also a 1:2 function of B and P. If the enzyme is saturated with A and Q, the expression for the velocity is

$$\dot{v}^{a\to q} = \cfrac{V_f/K_c}{\cfrac{\dot{K}_a K_{ip_1}K_{iq} + K_{ia}K_{ip_1}\dot{K}_q + K_{ia}K_{ip_2}\dot{K}_q}{K_{ia}K_{ip_1}K_{ip}} + \cfrac{\dot{K}_a K_{ib_2}}{K_{ia}(B)} + \cfrac{\dot{K}_a(B)}{K_{ia}K_{ib_1}}}{+ \cfrac{K_{ip_2}\dot{K}_q}{K_{iq}(P)} + \cfrac{\dot{K}_q(P)}{K_{ip_1}K_{iq}} + \cfrac{\dot{K}_a K_{ib_2}(P)}{K_{ia}K_{ip_2}} + \cfrac{K_{ip_2}\dot{K}_q(B)}{K_{ib_2}K_{iq}(P)}}$$

$$(8.6)$$

The reason for the requirement for B and P in the A → Q exchange as well as the reason why these reactants inhibit the exchange at high concentrations is apparent by considering the pathway shown in Figure 8.3. The interconversion of the labelled enzyme species must take place if isotopic exchange is to occur. The rapid inter-conversion of the EA and EAB complexes requires the presence of substrate B and yet is inhibited by high concentrations of B. Likewise, the rapid inter-conversion of EAB and FQ requires the presence of product P and yet is inhibited by high concentrations of P. For this reason, reactants which interact with the enzyme between the binding and dissociation of the exchangeable reactants are both required and inhibitory at high concentrations. Thus, isotopic exchange is a useful technique for determining binding orders.

Equation (8.7) is the rate equation for the Q → A exchange.

$$\dot{v}_{q \to a} = \left[k_{54}(\dot{Q}) \frac{N_{EA}}{N_t} \frac{(F)}{E_t} \right] E_t \qquad (8.7)$$

The N_{ea} path is the path which connects all the labelled species of the enzyme to the EA complex and also results in the dissociation of A from the enzyme. Thus it consists of $k_{21}k_{32}k_{43}(P)$. If the Q → A exchange is to take place, both Q and A must be present in the reaction medium initially. Equation (8.7) contains the concentration of Q, and by reference to Fig. 8.2, it can be seen that only the first two terms of $(F)/E_t$ contain the concentration of A, and these terms contain (A) (B). When proper substitutions are made in eq. (8.7), it is seen that the equations for the A → Q and Q → A exchanges are identical.

If there is an exchange between A and P, the equation for the A → P exchange is

$$\dot{v}_{a \to p} = \left[k_{12}(\dot{A}) \frac{N_{EAB}}{N_t} \frac{(E)}{E_t} \right] E_t \qquad (8.8)$$

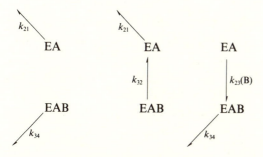

$$N_t = k_{21}[k_{32} + k_{34}] + k_{23}k_{34}(B)$$

Fig. 8.4. Paths involved in A → P and P → A exchange for the reaction sequence in Fig. 8.1. Only those enzyme species which become labelled with isotope are shown.

The paths between isotopically labelled enzyme species are shown in Figure 8.4. The first three terms in the distribution expression for the free enzyme contain the concentration of P, but the third term contains the concentrations of C and P. Since N_{eab} contains (B), the presence of C in the reaction medium will result in the net synthesis of P. Hence, only those terms in (E)/E_t which contain (P) (Q) will assure isotopic exchange rather than isotopic exchange plus net synthesis. It might seem surprising that Q is required for the A → P exchange since Q does not interact with a species of the enzyme which becomes labelled during the A → P exchange, but reference to Figure 8.1 shows the reason for the requirement for Q. In the absence of C, the reaction sequence will not proceed beyond the F species. Hence, in the absence of both C and Q, there would be no way of converting the enzyme in the form of F back to those species of the enzyme which participate in the exchange. Eventually, all the enzyme would exist as F and the exchange process would stop.

Multiplication of the numerator and denominator of eq. (8.8) by the factor $k_{45}k_{56}k_{61}$/coef. ABC gives rise to the following equation.

$$\dot{v}_{a \to p} = \frac{V_f(\dot{A})(B)(P)(Q)}{\dfrac{\text{constant} + \text{coef. B(B)}}{\text{coef. ABC} \times \text{coef. PQ}}} \dot{D} \tag{8.9}$$

The term $\dot{D}$ is identical to eq. (8.3). The rate equation for the A → P exchange will be a 1:2 function in (B), but not in the concentration of any other reactant. Thus, while B is required for the exchange to take place, high concentrations of B will be inhibitory, and this is apparent by an inspection of Figure 8.1.

8.2 Isotopic exchange in a random enzyme sequence

Isotopic exchange studies are particularly useful as a means of distinguishing between ordered and random reaction sequences. To demonstrate this fact, consider a reaction sequence similar to Figure 8.1, but in which either product P or Q may dissociate first from the enzyme.

The equation for the rate of the A → Q exchange is,

$$\dot{v}_{a \to q} = \left[k_{12}(\dot{A}) \frac{N_{FQ} + N_{EAB}}{N_t} \frac{(E)}{E_t} \right] E_t \tag{8.10}$$

The paths which the labelled enzyme species may follow are the following: Equation (8.10) contains the distribution expression for the free enzyme.

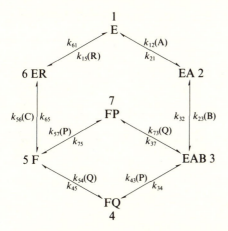

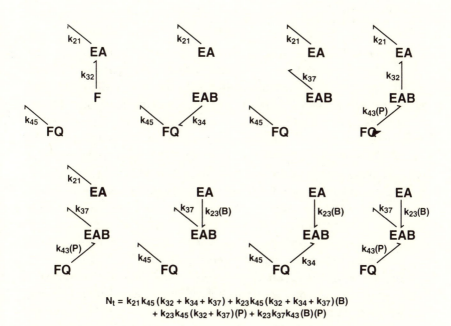

Fig. 8.5. Reaction sequence similar to Figure 8.1 except the dissociation of products is random rather than ordered.

$$N_t = k_{21}k_{45}(k_{32} + k_{34} + k_{37}) + k_{23}k_{45}(k_{32} + k_{34} + k_{37})(B)$$
$$+ k_{23}k_{45}(k_{32} + k_{37})(P) + k_{23}k_{37}k_{43}(B)(P)$$

Fig. 8.6. Paths involved in the A → Q and Q → A exchange for the reaction sequence in Figure 8.1. Only those enzyme species which become labelled with isotope are shown.

This expression contains 20 terms, and in vector form they are the following:

$$(E)/E_t$$

$(0, 1, 2, 5, 6, 1, 5)$ (C)
$(0, 1, 4, 5, 6, 1, 5)$ (C)
$(0, 1, 7, 3, 6, 1, 5)$ (C)

$\qquad\qquad\qquad (0, 3, 7, 3, 6, 1, 5)$ (B) (C) (P)

$(0, 3, 4, 5, 6, 1, 5)$ (B) (C)
$(0, 3, 7, 5, 6, 1, 5)$ (B) (C) $(0, 3, 4, 5, 6, 1, 3)$ (B) (C) (Q)
$(0, 1, 2, 3, 6, 1, 5)$ $(0, 1, 2, 3, 6, 1, 3)$ (C) (P) (Q)   (8.11)
$(0, 1, 7, 3, 6, 1, 5)$ (C) (P)

$\qquad\qquad\qquad (0, 1, 2, 3, 7, 1, 3)$ (P)^2 (Q)
$(0, 1, 2, 5, 6, 1, 3)$ (C) (Q) $(0, 1, 2, 3, 7, 5, 3)$ (P)^2 (Q)
$(0, 1, 4, 5, 6, 1, 3)$ (C) (Q)

$\qquad\qquad\qquad (0, 1, 2, 3, 4, 1, 3)$ (P) (Q)^2
$(0, 1, 2, 3, 4, 1, 5)$ (P) (Q) $(0, 1, 2, 3, 4, 5, 3)$ (P) (Q)^2
$(0, 1, 2, 3, 4, 5, 5)$ (P) (Q)
$(0, 1, 2, 3, 7, 1, 3)$ (P) (Q)
$(0, 1, 2, 5, 7, 5, 3)$ (P) (Q)

The reactants C and R are not required for the $A \rightarrow Q$ exchange and therefore the terms which contain either (C) or (R) can be ignored. When the proper terms are substituted into eq. (8.11), the expression will be complex, but it can be seen that the numerator of the equation will consist of terms which contain (P)^2 and (Q)^2. Hence, in contrast to eq. (8.5) for the ordered reaction sequence, the equation for the random reaction sequence will be a 2:2 function in the concentrations of P and Q. The possible shapes of a 2:2 rational polynomial are discussed in Chapter 12, but it is not difficult to visualize that neither product P nor Q is certain to inhibit the exchange. If one of the alternate paths is preferred kinetically, there may be some inhibition, but this will not be as drastic as that observed with a rate equation which is a 1:2 function. The reason is that if the concentration of P is sufficiently high to inhibit the conversion of the EAB complex to the FQ complex, the exchange can take place by conversion of the EAB complex to the FP species. Thus, isotopic exchange can be employed to obtain information which might be difficult or impossible to obtain by initial velocity and product inhibition studies. It should be recognized that high concentrations of B will inhibit the $A \rightarrow Q$ exchange in the sequence portrayed in Figure 8.5 because the binding of substrates is not random in the reaction sequence.

The equations derived here assume steady state, but chemical equilibrium is not required. Equations have been derived for isotopic exchange under equilibrium conditions[1-3]. The treatments which were developed for equilibrium conditions lead to conclusions which are identical to those developed by the steady state treatment. An advantage to the steady state treatment is that the variables can be varied independently. In the case of equilibrium, substrates and products have to be varied in pairs such that equilibrium is maintained.

8.3 Problems for Chapter 8

8.1) The following observations have been made in isotopic exchange experiments involving an enzyme which catalyzes the following reaction

$$A + B \rightleftharpoons P + Q + R.$$

The reaction has been shown to exhibit the kinetic behavior of a ping-pong mechanism. It has been found that an isotopic exchange will occur between substrate A and either product P or Q in the absence of B. In the absence of B, the A → P exchange has been shown to require the presence of Q. The A → Q exchange requires the presence of P, but high concentrations of P severely inhibit the A → Q exchange.

a) What is the most likely reaction sequence of this enzymic reaction?
b) Why is Q required for the A → P exchange?
c) Why is a random mechanism not likely?

References

1. Boyer, P. D. (1959). Uses and limitations of measurements of rates of isotopic exchange and incorporation in catalyzed reactions. *Arch. Biochem. Biophys.* **82**: 387–410.
2. Morales, M. F., Horovitz, M. and Botts, J. (1962). The distribution of tracer substrate in an enzyme-substrate system at equilibrium. *Arch. Biochem. Biophys.* **99**: 258–64.
3. Boyer, P. D. and Silverstein, E. (1963). Equilibrium reaction rates and enzyme mechanisms. *Acta Chem. Scand.* **17**: s195–s202.
4. Alberty, R. A., Bloomfield, V., Peller, L. and King, E. L. (1962). Multiple intermediates in steady-state enzyme kinetics. IV. The steady state kinetics of isotopic exchange in enzyme-catalyzed reactions. *J. Am. Chem. Soc.* **84**: 4381–84.
5. Cleland, W. W. (1967). Enzyme kinetics. *Ann. Rev. Biochem* **36**: 77–112.
6. Schulz, A. R. and Fisher, D. D. (1970). Computer-based derivation of rate equations for enzyme-catalyzed reactions. II. Rate equations for isotopic exchange. *Canad. J. Biochem.* **48**: 922–34.

9

Kinetic isotope effect on steady state parameters

In the previous chapter, the enzyme-catalyzed isotopic exchange was considered as a means of obtaining information concerning the reaction sequence of an enzymic reaction. In the present chapter, the effect of isotope mass on the rate constants of those steps which involve the formation or cleavage of a covalent bond will be considered with respect to its effect on the steady state parameters. The basis for an isotope effect on the rate constants in which covalent bonds are formed or broken is discussed in detail elsewhere[1,2], and those discussions go beyond the scope of this book.

9.1 The basis for the kinetic isotope effect on rate constants

For the purpose of this treatment, it is sufficient to state that substitution of the normal atom in a covalent bond with an atom of greater mass will always favor formation of a stronger bond, and that will result in a decrease in the rate constants involved in the formation or rupture of that bond. For example, the rate constant for the non-enzymic formation or rupture of a R—H bond is approximately 15 times greater than the rate constant for the formation or rupture of a R—D bond. In the foregoing, H represents a hydrogen atom while D represents a deuterium atom. In like manner, the rate constant for the formation or rupture of a R—H bond is approximately 50 times greater than that for a R—T bond. If the step in which a covalent bond is formed or ruptured were very much the rate limiting step in an enzymic reaction, one might expect the difference in rate constant to be reflected in a comparable difference in maximal velocity. Hence, the V_f^h/V_f^d ratio should be 15, where V_f^h is the maximal velocity for the substrate with a R—H bond and V_f^d is the maximal velocity for the substrate with a R—D bond. The ratios observed with many enzymes are much less than the predicted value[3-5]. As a matter of fact, ratios in the range of 4 to 8 are not uncommon. Northrop[3] developed the following treatment to explain this phenomenon.

9.2 Use of the kinetic isotope effect in steady state enzyme kinetic studies

Consider Figure 9.1 as a model for an ordered, sequential bi-bi enzymic reaction in the absence of the products. The sequence includes both an EAB and an EPQ ternary complex. The steps which involve the formation and rupture of a covalent bond intervene between these complexes, and the relative rate of these steps is of interest in this treatment. For the purpose of this discussion the concentrations of the products will be assumed to be equal to zero, and therefore, the steps at which the products dissociate are visualized as irreversible. Figure 9.2 shows the enzyme distribution

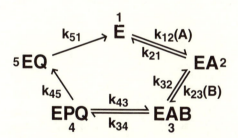

Fig. 9.1. Enzyme reaction sequence of and ordered, bi-bi, sequential reaction which contains two ternary complexes. The concentration of both products is assumed to be equal to zero.

$(E)/E_t$	$(EA)/E_t$	$(EAB)/E_t$	$(EPQ)/E_t$	$(EQ)/E_t$
$(0,1,2,3,1)$	$(2,0,2,3,1)(A)$	$(2,3,0,3,1)(A)(B)$	$(2,3,4,0,1)(A)(B)$	$(2,3,4,5,0)(A)(B)$
$(0,1,2,5,1)$	$(2,0,2,5,1)(A)$	$(2,3,0,5,1)(A)(B)$		
$(0,1,4,5,1)$	$(2,0,4,5,1)(A)$			
$(0,3,4,5,1)(B)$				

Fig. 9.2. Enzyme distribution expressions for the reaction sequence in Fig. 9.1.

expressions for the reaction under consideration. The relevant steady state parameters are,

$$
V_f = \frac{(2,3,4,5,1)E_t}{\begin{array}{l}(2,3,4,5,0)\\(2,3,4,0,1)\\(2,3,0,3,1)\\(2,3,0,5,1)\end{array}} = \frac{k_{34}k_{45}k_{51}E_t}{k_{34}k_{45}+k_{34}k_{51}+k_{43}k_{51}+k_{45}k_{51}}
\tag{9.1}
$$

$$
K_a = \frac{(0,3,4,5,1)}{\begin{array}{l}(2,3,4,5,0)\\(2,3,4,0,1)\\(2,3,0,3,1)\\(2,3,0,5,1)\end{array}} = \frac{k_{34}k_{45}k_{51}}{k_{12}(k_{34}k_{45}+k_{34}k_{51}+k_{43}k_{51}+k_{45}k_{51})}
\tag{9.2}
$$

$$
K_b = \frac{\begin{array}{l}(2,0,2,3,1)\\(2,0,2,5,1)\\(2,0,4,5,1)\end{array}}{\begin{array}{l}(2,3,4,5,0)\\(2,3,4,0,1)\\(2,3,0,3,1)\\(2,3,0,5,1)\end{array}} = \frac{k_{51}(k_{32}k_{43}+k_{32}k_{45}+k_{34}k_{45})}{k_{23}(k_{34}k_{45}+k_{34}k_{51}+k_{43}k_{51}+k_{45}k_{51})}
\tag{9.3}
$$

$$
\frac{V_f}{K_a} = k_{12}E_t
\tag{9.4}
$$

$$
\frac{V_f}{K_b} = \frac{k_{23}k_{34}k_{45}E_t}{k_{32}k_{32}+k_{32}k_{43}+k_{34}k_{45}}
\tag{9.5}
$$

The rate constants for the steps in Fig. 9.1 which involve the formation and rupture of a covalent bond are k_{34} and k_{43}. These are the only rate constants which would be expected to exhibit an isotope effect. If the covalent bond involved were of the nature R—H in the treatment to follow, the rate constants will be identified as k_{34}^h, k_{43}^h if the experiment were conducted with a substrate with a hydrogen atom. In like manner, the rate constants will be identified as k_{34}^d, k_{43}^d if the experiments were conducted with the substrate containing a deuterium atom.

   In order to compare the maximal velocity of the reaction in the presence of the hydrogen-substrate with that of the deuterium-substrate, it is convenient to divide the numerator and denominator of the right-hand side of eq. (9.1) by the denominator term which does not contain a rate constant for the catalytic step. The equation becomes

$$V_f = \frac{k_{34} E_t}{\dfrac{k_{34} + k_{43}}{k_{45}} + \dfrac{k_{34}}{k_{51}} + 1} \tag{9.6}$$

The ratio of maximal velocity in the presence of the hydrogen-substrate to that of the deuterium-substrate is given by eq. (9.7).

$$\frac{V_f^h}{V_f^d} = \frac{k_{34}^h}{k_{34}^d} \left[\frac{\dfrac{k_{34}^d + k_{43}^d}{k_{45}} + \dfrac{k_{34}^d}{k_{51}} + 1}{\dfrac{k_{34}^h + k_{43}^h}{k_{45}} + \dfrac{k_{34}^h}{k_{51}} + 1} \right] \tag{9.7}$$

If the steps at which the covalent bond is formed or ruptured were very much slower than the other steps represented in eq. (9.7), that is, if k_{34} and $k_{43} \ll k_{45}$ and k_{51}, eq. (9.7) would become

$$\frac{V_f^h}{V_f^d} \cong \frac{k_{34}^h}{k_{34}^d} \cong 15 \tag{9.8}$$

The predicted ratio is observed under this condition. However, if the relevant rate constants were essentially equal, that is, if $k_{34} = k_{43} = k_{45} = k_{51}$, the following ratio would be observed:

$$\frac{V_f^h}{V_f^d} = 15 \left[\frac{\dfrac{2}{15} + \dfrac{1}{15} + 1}{2 + 1 + 1} \right] = 4.5 \tag{9.9}$$

This ratio is closer to that observed with several of the enzymes which have been investigated in this manner. This provides direct evidence that the Briggs-Haldane treatment is more appropriate than the assumption on which the Henri and Michaelis-Menten treatment were based.

Information about the relative magnitude of additional rate constants in the reaction sequence portrayed in Figure 9.1 can be obtained. Division of the numerator and denominator of the right-hand side of eq. (9.5) by $k_{32} k_{45}$ gives

$$\frac{V_f}{K_b} = \frac{\dfrac{k_{23} k_{34}}{k_{32}} E_t}{\dfrac{k_{34}}{k_{32}} + \dfrac{k_{43}}{k_{45}} + 1} \tag{9.10}$$

Equation (9.11) is an expression of the ratio of this parameter for the hydrogen-substrate compared to the deuterium-substrate.

$$\frac{(V_f/K_b)^h}{(V_f/K_b)^d} = \frac{k_{34}^h}{k_{34}^d} \left[\frac{\dfrac{k_{34}^d}{k_{32}} + \dfrac{k_{43}^d}{k_{45}} + 1}{\dfrac{k_{34}^h}{k_{32}} + \dfrac{k_{43}^h}{k_{45}} + 1} \right] \tag{9.11}$$

If k_{34} and $k_{43} \ll k_{32}$ and k_{45}, it can be easily calculated that the ratio will be approximately 15. However, if these four rate constants were equal

$$\frac{(V_f/K_b)^h}{(V_f/K_b)^d} = 15 \left[\frac{\frac{1}{15} + \frac{1}{15} + 1}{1 + 1 + 1} \right] = 5.67 \tag{9.12}$$

Once again, the ratio is much less than the expected value, but closer to the value obtained for several enzymes that have been investigated in this manner. If the enzyme-catalyzed reaction is reversible, similar experiments could be conducted using products P and Q as substrates and in this manner estimates of the relative magnitude of the values of k_{34} and k_{43} compared to k_{21}, k_{23} and k_{54} could be obtained.

As stated earlier, studies of the steady state kinetic behavior of enzymes usually do not provide information about the magnitude of individual rate constants. However, the kinetic isotope effect provides a means by which studies of the kinetic isotope effect can be exploited to provide information about the relative magnitude of the rate constants associated with the formation or rupture of a covalent bond compared to the other rate constants in the sequence. This procedure is most useful if the covalent bond involved is a R—H bond because the difference in atomic mass of the hydrogen atom and its isotopes is much greater than other atoms.

9.3 Problems for Chapter 9

9.1   Calculate the V_f^h/V_f^d ratio for the reaction sequence in Figure 9.1 if $k_{45} = k_{51} = 10 \times k_{34}$ and $k_{34} = k_{43}$ and $k_{34}^h/k_{34}^d = 15$.

9.2   Calculate the V_f^h/V_f^d ratio for the reaction sequence in Figure 9.1 if $k_{45} = k_{51} = 10 \times k_{34}$ and $k_{43} = 0$ and $k_{34}^h/k_{34}^d = 15$.

9.3   Derive the expressions for V_f^h/V_f^d and $(V_f/K_a)^h/(V_f/K_a)^d$ for the reaction sequence in Figure 9.3.
      Assume that the concentrations of both products are equal to zero.

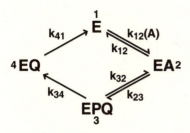

Fig. 9.3. Reactions sequence of an ordered uni-bi, sequential enzymic reaction. The concentration of both products is assumed to be equal to zero.

References

1. Jencks, W. P. (1969). *Catalysis in Chemistry and Enzymology*, pp. 243–81, New York, McGraw-Hill.
2. Laidler, K. J. and Bunting, P. S. (1973). *The Chemical Kinetics of Enzyme Action*, 2d. Ed., pp. 233–53, Oxford, Oxford University Press.
3. Northrop, D. B. (1975) Steady-state analysis of kinetic isotope effects in enzymic reactions. *Biochem.* **14**: 2644–51.
4. Schimerlik, M. I., Grimshaw, C. E. and Cleland, W. W. (1977). Determination of the rate-limiting steps for malic enzyme by the use of isotope effects and other kinetic studies. *Biochem.* **16**: 571–76.
5. Cleland, W. W., O'Leary, M. H. and Northrop, D. B. (1977). *Isotope Effects on Enzyme-catalyzed Reactions*, Baltimore, University Park Press.

10

The effect of pH on enzyme kinetics

One of the many aspects of biochemistry to which L. Michaelis made a significant contribution was the effect of pH on the dissociation of polyvalent acids and of proteins[1]. It was noted early in the study of enzyme action that enzymes generally exhibited a bell-shape curve if enzymic activity was plotted against pH. An analogous plot is observed if one plots the concentration of the zwitterion species of a monoamino, monocarboxylic acid against pH. Because of the simplicity of the amino acid, the discussion of the effect of pH will start with this model.

10.1 Michaelis pH functions of a simple amino acid

In an aqueous solution, a monoamino, monocarboxylic acid is distributed between three species, namely the fully protonated acid, the zwitterion and the fully dissociated base are shown in eq. 10.1 and Figure 10.1.

$$A_t = (H_2A^+) + (HA^0) + (A^-) \tag{10.1}$$

The superscript in eq. (10.1) indicates the net charge of each species of the acid. The protons dissociate from the acidic groups according to the following:

$$H_aA^+ \rightleftharpoons HA^0 + H^+, \quad K_1 = \frac{(HA^0)(H^+)}{(H_2A^+)} \tag{10.2}$$

$$HA^0 \rightleftharpoons A^- + H^+, \quad K_2 = \frac{(A^-)(H^+)}{(HA^0)} \tag{10.3}$$

The following relationships follow:

$$(H_2A^+) = \frac{(HA^0)(H^+)}{K_1} = \frac{(A^-)(H^+)^2}{K_1 K_2} \tag{10.4}$$

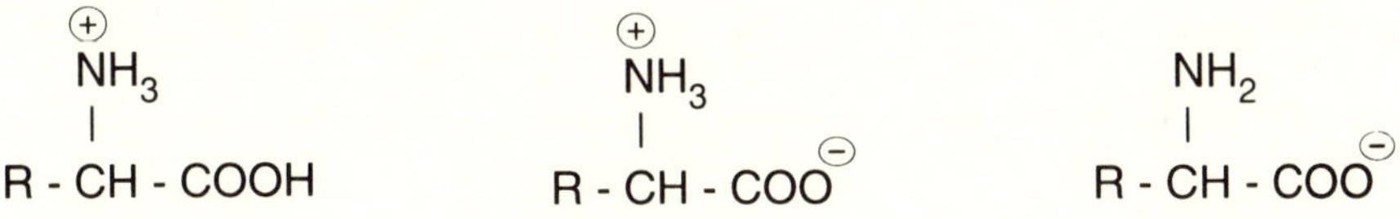

$$\overset{\oplus}{NH_3}$$
$$|$$
$$R\text{-}CH\text{-}COOH$$

$$\overset{\oplus}{NH_3}$$
$$|$$
$$R\text{-}CH\text{-}COO^{\ominus}$$

$$NH_2$$
$$|$$
$$R\text{-}CH\text{-}COO^{\ominus}$$

$$H_2A^+ \qquad\qquad\qquad HA^o \qquad\qquad\qquad A^-$$

Fig. 10.1. Ionic species of a monoamino, monocarboxylic acid.

$$(HA^0) = \frac{K_1(H_2A)}{(H^+)} = \frac{(A^-)(H^+)}{K_2} \tag{10.5}$$

$$(A^-) = \frac{K_2(HA^0)}{(H^+)} = \frac{K_1 K_2(H_2A^+)}{(H^+)^2} \tag{10.6}$$

Equation (10.1) can be expressed in terms of A_t and any one of the species.

$$A_t = (H_2A^+)\left[1 + \frac{K_1}{(H^+)} + \frac{K_1 K_2}{(H^+)^2}\right] = (H_2A^+)f^+ \tag{10.7}$$

$$A_t = (HA^0)\left[1 + \frac{(H^+)}{K_1} + \frac{K_2}{(H^+)}\right] = (HA^0)f^0 \tag{10.8}$$

$$A_t = (A^-)\left[1 + \frac{(H^+)}{K_2} + \frac{(H^+)^2}{K_1 K_2}\right] = (A^-)f^- \tag{10.9}$$

The expression in brackets in the preceding equations is abbreviated f, and is called the Michaelis pH function. It is readily recognized that the reciprocal of the Michaelis pH function is the fraction of the acid present as any given species of the acid. If alanine is taken as a typical monoamino, monocarboxylic acid, the dissociation constant of the carboxyl group, K_1, is 4.49×10^{-3} and the dissociation constant of the protonated amino group, K_2, is 1.36×10^{-10}. The fraction of the total acid present as each species acid as a function of pH is shown in Figure 10.2.

It is convenient to express the Michaelis pH function for the zwitterion in the logarithmic form.

$$pf^0 = -\log f^0 = -\log\left[1 + \frac{(H^+)}{K_1} + \frac{K_2}{(H^+)}\right] \tag{10.10}$$

At a pH of 1, $((H^+)/K_1) \gg 1 + (K_2/(H^+))$ and eq. (10.10) becomes

$$pf^0 \cong -\log\frac{(H^+)}{K_1} = pH - pK_1 \tag{10.11}$$

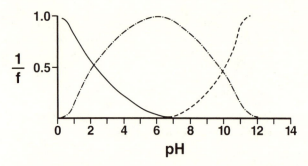

Fig. 10.2. Plots of the fraction of a monoamino, monocarboxylic acid as the fully protonated acid _____, as the neutral zwitterion _._._, and as the free base _____.

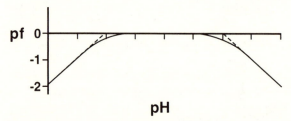

Fig. 10.3. Plot of pf^0 of the zwitterion form of a monoamino, monocarboxylic acid as a function of pH. The negative logarithm of the dissociation constant of the carboxyl group is pK_1 while pK_2 is the negative logarithm of the dissociation constant of the amino group.

where $pK_1 = -\log K_1$. Thus in this region a plot of pf^0 versus pH is a straight line with slope equal to 1. When $(H^+) = K_1$, $K_2/(H^+)$ is still very small so $pf^0 \cong -\log(2) = -0.301$. The peak value of $1/f^0$ in Figure 10.1 is 1. The value of pf^0 at that point is zero. At a still higher pH where $pH = pK_2$, $(H^+)/K_1$ is negligible so $pf^0 \cong -\log(2) = -0.301$. Finally, at a pH of 11 or above, eq. (10.10) becomes

$$pf^0 \cong -\log \frac{K_2}{(H^+)} = -pH + pK_2 \qquad (10.12)$$

The foregoing is portrayed in Figure 10.3. The point at which the extrapolated straight line intersects the pH axis is equal to the pK values because the slope of these lines is either 1 or -1 and the points at which these lines intersect the pf axis is $-pK_1$ and pK_2.

In the foregoing discussion, it has been assumed that the two dissociation constants are sufficiently different so that the more basic group never dissociates before the more acidic group. The situation where the dissociation constants are close to one another has been discussed elsewhere[3,4].

10.2 Michaelis pH functions of an enzyme

The bell-shaped curve which shows the fraction of amino acid present as the zwitterion is similar to the curves obtained when the velocity of an enzyme-catalyzed reaction is plotted against pH. This led to the conclusion that intermediate ionized species of an enzyme reaction sequence are involved in the reaction sequence[1-3,5-8]. This is portrayed in Figure 10.4. The assumption that the dissociation and association of a proton is very rapid as compared to any of the steps in the catalytic cycle will be made such that the acid-base reactions are at equilibrium. This assumption is not mandatory to the derivation of an equation, but simplifies the derivation considerably and the conclusions are consistent with those in which this assumption has not been made[5].

For any rate constant, if k is the observed pH dependent rate constant, then $k = \tilde{k} f^n$ where $\tilde{k}$ is the pH independent rate constant and f^n is the appropriate Michaelis pH function[5]. The foregoing follows from eqs. (10.7)–(10.9), for the pH dependent rate constant applies to the total concentration of the species involved in the reaction. Figure 10.5 presents the enzyme distribution expressions for the reaction model in Fig. 10.4, but they are expressed in terms of pH independent steady state parameters and Michaelis pH functions. As has been done throughout this textbook, all the

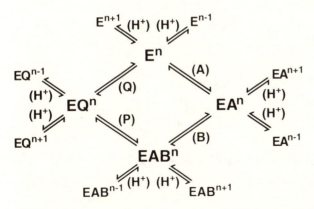

Fig. 10.4. The ionic species of an ordered, bi-bi, sequential enzyme reaction sequence. The active species are those with charge equal to n. The actual charge on any species is not specified.

$(E)/E_t$	$(EA)/E_t$	$(EAB)/E_t$	$(EQ)/E_t$
$\tilde{K}_{ia}\tilde{K}_b f_e^n$ $\tilde{K}_a f_e^n(B)$	$\tilde{K}_b f_{ea}^n(A)$	$f_{eab}^n(A)(B)$	$f_{eq}^n(A)(B)$
$\dfrac{\tilde{K}_{ia}\tilde{K}_b}{\tilde{K}_{ip_1}} f_e^n(P)$	$\dfrac{\tilde{K}_b}{\tilde{K}_{ip_1}} f_{ea}^n(A)(P)$	$\dfrac{f_{eab}^n}{\tilde{K}_{ip_2}}(A)(B)(P)$	$\dfrac{\tilde{K}_b f_{eq}^n}{\tilde{K}_{iq}}(B)(Q)$
	$\dfrac{\tilde{K}_{ia}\tilde{K}_b}{\tilde{K}_p\tilde{K}_{iq}} f_{ea}^n(P)(Q)$	$\dfrac{\tilde{K}_{ia}\tilde{K}_b}{\tilde{K}_p\tilde{K}_{iq}} f_{eab}^n(P)(Q)$	$\dfrac{\tilde{K}_{ia}\tilde{K}_b}{\tilde{K}_{iq}} f_{eq}^n(Q)$
		$\dfrac{\tilde{K}_a f_{eab}^n}{\tilde{K}_{ip_3}\tilde{K}_{iq}}(B)(P)(Q)$	

Fig. 10.5. Enzyme distribution expressions for the reaction sequence in Fig. 10.4. The distribution expressions are stated in terms of pH independent kinetic parameters and the Michaelis pH functions of the enzyme species.

terms in the numerator and denominator of the rate equation are divided by the coefficient of the denominator term which contains the concentrations of the substrates to the highest equal power. The complete rate equation for the model shown in Figure 10.4 is

$$
v = \cfrac{\dfrac{\tilde{V}_f}{f_{eab}^n + f_{eq}^n}\left(1 - \dfrac{\Gamma}{\tilde{K}_{eq}}\right)}{\begin{aligned} &1 + \dfrac{\tilde{K}_a}{(A)}\left(\dfrac{f_e^n}{f_{eab}^n + f_{eq}^n}\right) + \dfrac{\tilde{K}_b}{(B)}\left(\dfrac{f_{eq}^n}{f_{eab}^n + f_{eq}^n}\right) \\ &+ \dfrac{\tilde{K}_{ia}\tilde{K}_b}{(A)(B)}\left(\dfrac{f_e^n}{f_{eab}^n + f_{eq}^n}\right) + \dfrac{(P)}{\tilde{K}_{ip_1}}\left(\dfrac{f_{eab}^n}{f_{eab}^n + f_{eq}^n}\right) \\ &+ \dfrac{\tilde{K}_b(P)}{\tilde{K}_{ip_1}(B)}\left(\dfrac{f_{eq}^n}{f_{eab}^n + f_{eq}^n}\right) + \dfrac{\tilde{K}_{ia}\tilde{K}_b(P)}{\tilde{K}_{ip_1}(A)(B)} + \left(\dfrac{f_e^n}{f_{eab}^n + f_{eq}^n}\right) \\ &+ \dfrac{\tilde{K}_a(Q)}{\tilde{K}_{iq}^n(A)}\left(\dfrac{f_{eq}^n}{f_{eab}^n + f_{eq}^n}\right) + \dfrac{\tilde{K}_{ia}\tilde{K}_b(Q)}{\tilde{K}_{iq}(A)(B)}\left(\dfrac{f_{eq}^n}{f_{eab}^n + f_{eq}^n}\right) \\ &+ \dfrac{\tilde{K}_a(P)(Q)}{\tilde{K}_{ip_3}\tilde{K}_{iq}(A)}\left(\dfrac{f_{eab}^n}{f_{eab}^n + f_{eq}^n}\right) + \dfrac{\tilde{K}_{ia}\tilde{K}_b(P)(Q)}{\tilde{K}_p\tilde{K}_{iq}(A)(B)}\left(\dfrac{f_{eq}^n + f_{eab}^n}{f_{eab}^n + f_{eq}^n}\right) \end{aligned}} \tag{10.13}
$$

10.3 The effect of pH on steady state enzymic parameters

A comparison of eq. (10.13) with the equation expressed in terms of pH dependent parameters, namely eq. (6.1), shows the following relationships.

$$
V_f = \frac{\tilde{V}_f}{f_{eab}^n + f_{eq}^n} \tag{10.14}
$$

$$K_a = \tilde{K}_a \left(\frac{f_e^n}{f_{eab}^n + f_{eq}^n} \right) \tag{10.15}$$

$$K_b = \tilde{K}_b \left(\frac{f_{eq}^n}{f_{eab}^n + f_{eq}^n} \right) \tag{10.16}$$

$$K_{ia} = \tilde{K}_{ia} \left(\frac{f_e^n}{f_{eq}^n} \right) \tag{10.17}$$

$$K_{ip_1} = \tilde{K}_{ip_1} \tag{10.18}$$

$$K_{ip_2} = \tilde{K}_{ip_2} \left(\frac{f_{eab}^n + f_e^n}{f_{eab}^n} \right) \tag{10.19}$$

$$K_{iq} = \tilde{K}_{iq} \left(\frac{f_e^n}{f_{eq}^n} \right) \tag{10.20}$$

$$\frac{V_f}{K_a} = \frac{\tilde{V}_f}{\tilde{K}_a} \left(\frac{1}{f_e^n} \right) \tag{10.21}$$

$$\frac{V_f}{K_b} = \frac{\tilde{V}_f}{\tilde{K}_b} \left(\frac{1}{f_{ea}^n} \right) \tag{10.22}$$

$$V_f K_{ip_2} = \tilde{V}_f \tilde{K}_{ip_2} \left(\frac{1}{f_{eab}^n} \right) \tag{10.23}$$

$$\frac{V_f K_{iq}}{K_a} = \frac{\tilde{V}_f \tilde{K}_{iq}}{\tilde{K}_a} \left(\frac{1}{f_{eq}^n} \right). \tag{10.24}$$

For the reaction sequence under consideration, K_{ip_1} is independent of pH. However, the remaining steady state parameters defined in eqs. (10.14) through (10.20) are rather complex relationships of Michaelis pH functions. In contrast, the relationships expressed in eqs. (10.21) through (10.24) are each a function of a single Michaelis pH function, and this fact was recognized by Laidler[8]. By a judicious choice of the parameter plotted against pH, it is possible to analyze the effect of pH on each enzyme species in the catalytic sequence. For example, eqs. (10.21) and (10.22) could be re-written as

$$\log \left(\frac{V_f}{K_a} \right) = \log \left(\frac{\tilde{V}_f}{\tilde{K}_a} \right) - \log \left[1 + \frac{(H^+)}{K_{e_1}} + \frac{K_{e_2}}{(H^+)} \right] \tag{10.25}$$

$$\log \left(\frac{V_f}{K_b} \right) = \log \left(\frac{\tilde{V}_f}{\tilde{K}_b} \right) - \log \left[1 + \frac{(H^+)}{K_{ea_1}} + \frac{K_{ea_2}}{(H^+)} \right] \tag{10.26}$$

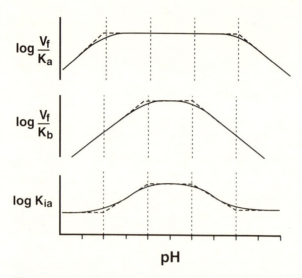

Fig. 10.6. Hypothetical plots of logarithmic parameters as a function of pH for the enzymic sequence in Fig. 10.3. The plot of log $V_f \backslash Ka$ versus. pH shows the pK values for the free enzyme. The plot of log V_f/K_b versus pH shows the pK values for the EA complex. The plot of log K_{ia} versus pH is a composite of the pK values for the free enzyme and the EA complex.

Dependent on the hydrogen ion concentration, the term in brackets in eqs. (10.25) and (10.26), that is, the Michaelis pH functions, could be expressed as eqs. (10.11) or (10.12) and the following plots might be obtained. Figure 10.6 portrays a situation where $K_{e_1} < K_{ea_1}$ and $K_{e_2} > K_{ea_2}$. The plot of log K_{ia} versus pH is the plot of log V_f/K_b versus pH minus the plot of log V_f/K_a versus pH. Since all the foregoing figures represent hypothetical plots, the stability of the enzyme as a function of pH presents no problem. In any actual experiment, it would be rare for any enzyme to exhibit this degree of stability. A more common observation is that only one portion of the foregoing plots can be evaluated.

10.4 Effect of substrate ionization on steady state parameters

The treatment of the effect of pH on the kinetic behavior of enzymes to this point has ignored any effect that pH might exert on the substrate. However, a substrate may contain one or more acidic groups, and if such is the case, the enzyme might be specific for one of the ionic species or, all the ionic species may serve as the substrate. For example, in the enzyme model under consideration, substrate A might be a monoamino-monocarboxylic acid. The enzyme distribution expressions given in Figure 10.4 express the

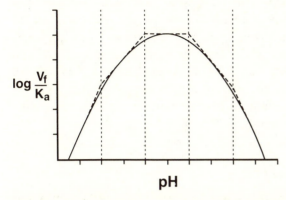

Fig. 10.7. Hypothetical plot of the log V_f/K_a versus pH for the reaction sequence in Figure 10.3 when substrate A is an amino acid and the enzyme is specific for the neutral zwitterion.

substrate concentrations in terms of the total substrate, that is $(A) = A_t$. If the enzyme were to exhibit no preference with regard to the ionic species of substrate A, the concentration of the actual substrate would be equal to the total substrate concentration. On the other hand, if the enzyme were specific for one ionic species of substrate A, A_t would have to be expressed in terms of eqs. (10.7), (10.8) or (10.9). Thus, if the enzyme were specific for the zwitterion, then eq. (10.21) would become

$$\frac{V_f}{K_a} = \frac{\tilde{V}_f}{\tilde{K}_a}\left(\frac{1}{f_e^n}\right)\left(\frac{1}{f^0}\right) \tag{10.27}$$

A plot of log V_f/K_a versus pH would appear as in Figure 10.7 if the substrate concentration were A_t. As predicted by eq. (10.27), the initial slope of the extrapolated line in Figure 10.7 is 2. The second linear segment has a slope of 1. If substrate concentrations were calculated in terms of the zwitterion concentration, that is, the ionic species which would be recognized by the enzyme, the Michaelis constant for substrate A would be decreased and the plot of log V_f/K_a versus pH would exhibit the ionizations of the free enzyme only.

The presence of an ionizable group on a substrate somewhat complicates the analysis of the effect of pH. However, it is a complication which can be circumvented rather simply. The treatment presented in this chapter is based on the assumption that the various species of the enzyme exist in one of three states of dissociation with the intermediate species being the active species. This is an assumption which has been made in all treatments of the effect of pH on enzyme activity. Therefore, a slope in excess of 1 indicates that the dissociation of a group on the substrate is involved. The ionization constant(s) for a substrate should be known, and therefore, it is easy to calculate the concentration of each ionic species. Thus, it is usually possible to determine which form of the substrate is the true substrate for the enzyme. Once again, it is necessary for the enzyme kineticist to consider all the information available, to "ask" the enzyme the proper questions and then to interpret the information in a judicious manner in order to solve the mystery of the kinetic behavior of the enzyme.

10.5 Problems for chapter 10

10.1 The following data were obtained when substrate- saturation experiments were conducted using the same substrate with two different pH values.

(A) μM	velocity at pH 8.2 μmoles/minute	velocity at pH 8.7 μmoles/minute
1.43		0.0168
1.67		0.0194
2.00	0.0093	0.0227
2.50	0.0115	0.0275
3.33	0.0150	0.0375
5.00	0.0214	0.0476
10.00	0.0366	0.0740

Plot the data from both experiments on the same sheet of graph paper using any one of the linear plots. Determine the apparent maximum velocities and the apparent Michaelis constants for each experiment.

10.2 Substrate A in the experiments described in problem 10.1 is an acid which has a pK_a value of 9.37. With this information and the Henderson-Hasselbach equation, calculate the concentration of the basic form of substrate A in each experiment and plot the data from both experiments on a second sheet of graph paper. On the basis of these experiments, what is the effect of pH on the enzymic reaction under these conditions?

10.3 Derive the complete rate equation in the kinetic form in terms of pH-independent steady state parameters of the appropriate Michaelis pH functions for the model shown in Figure 10.8. Assume that the substrates and products do not contain acidic groups.

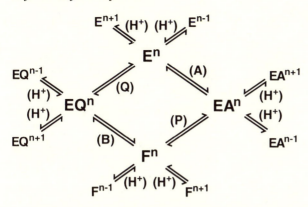

Fig. 10.8. The ionic species of a ping-pong enzyme reaction sequence.

References

1. Michaelis, L. and Davidsohn, H. (1911). Die wirkumg der wasserstoffonen auf das invertin. *Biochem. Z.* **35**: 386–412.
2. Dixon, M. (1953). The effect of pH on the affinities of enzymes for substrates and inhibitors. *Biochem. J.* **55**: 161–70.
3. Dixon, M. and Webb, E. C. (1979). Enzymes, 3rd. Ed., pp. 138–64, New York, Academic Press.
4. Adams, E. Q. (1916). Relations between the constants of dibasic acids and of amphoteric electrolytes. *J. Am. Chem. Soc.* **38**: 1503–10.
5. Waley, S. G. (1953). Some aspects of the kinetics of enzyme reactions. *Biochim. Biophys. Acta* **10**: 27–34.
6. Laidler, K. J. (1955). The influence of pH on the rate of enzyme reactions. Part 1. General theory. Trans. *Faraday Soc.* **51**: 528–39.
7. Laidler, K. J. (1955). The influence of pH on the rate of enzyme reactions. Part 2. The nature of the enzyme-substrate interactions. Trans. *Faraday Soc.* **51**: 540–49.
8. Laidler, K. J. (1955). The influence of pH on the rate of enzyme reactions. Part 3. Analysis of experimental results for various enzyme systems. Trans. *Faraday Soc.* **51**: 550–61.

Part Three

Non-hyperbolic enzyme kinetics

11

The causes of non-hyperbolic enzyme kinetics

Most of the enzymic reaction sequences which have been considered to this point have given rise to rate equations which describe rectangular hyperbolas. That is, the rate equations fall into a class of mathematical expressions called rational polynomials of the order 1:1. However, those reaction sequences which were characterized by substrate inhibition gave rise to rate equations which did not describe rectangular hyperbolas, but which were rational polynomials of order 1:2. Thus, in the latter cases, the rate equation consisted of a numerator which contained the concentration of the substrate to the first power while the denominator contained the concentration of the substrate to the second power. A slight modification of reaction sequences which have already been discussed can give rise to rate equations which are rational polynomials of even higher powers. The majority of the enzymes which exhibit non-hyperbolic kinetic behavior are allosteric enzymes or polymeric enzymes in which the subunit interactions exhibit either positive or negative cooperativity. Subunit interactions will be discussed in chapter 13. In the case of allosteric enzymes, the enzyme contains an allosteric site which is distinct from the active site. It must be understood that kinetic studies do not offer proof of allosterism. Conclusive evidence of a separate allosteric site must be obtained in studies of the physical structure of the enzyme.

11.1 Random enzyme reaction sequences

One explanation for an ordered binding of substrates to an enzyme is that the binding of the first substrate to the enzyme induces a conformational change in the enzyme such that the active site of the enzyme is in a proper configuration to allow the second substrate to bind properly[1]. As mentioned in chapter 1, there is evidence that the binding of a substrate to an

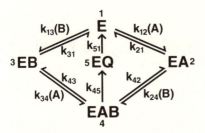

Fig. 11.1. Reaction sequence of a bi-bi, sequential model which is random with respect to substrates but ordered with respect to products. The concentration of both products is assumed to be zero.

enzyme induces a conformational change in the enzyme. While this explanation can provide an explanation for a *preferred* binding order, if a given conformation of the enzyme is possible in the presence of the substrate, from thermodynamic considerations that conformation should also be *possible* in the absence of the substrate, and for that reason a random binding should be possible. It may be that the required conformation which would allow the binding of the second substrate in the absence of the first substrate is so unlikely that the conformation required for the binding of the second substrate essentially never exists. Hence, if the substrate induced conformational change is the proposed explanation for an ordered sequence, a random binding order is possible, but perhaps not probable. There are, of course, other explanations for an ordered reaction sequence. For example, the first substrate to bind to the enzyme may provide a portion of the binding site for the second substrate. Glutamate dehydrogenase is an example of the latter mechanism[2].

A reaction sequence with a random binding is presented in Figure 11.1. The enzyme model portrayed in Figure 11.1 is a random substrate, ordered product, sequential, bi-bi mechanism. For convenience, both products are assumed to be absent so that the steps at which the products dissociate from the enzyme are irreversible. The following are the connection matrix and the Q matrix for the reaction sequence.

$$
U = \begin{vmatrix} 0 & A & B & 0 & 0 \\ 1 & 0 & 0 & B & 0 \\ 1 & 0 & 0 & A & 0 \\ 0 & 1 & 1 & 0 & 1 \\ 1 & 0 & 0 & 0 & 0 \end{vmatrix}, \quad Q = \begin{vmatrix} 2 & 3 & 4 \\ 1 & 4 & 0 \\ 1 & 4 & 0 \\ 2 & 3 & 5 \\ 1 & 0 & 0 \end{vmatrix}
$$

$(E)/E_t$	$(EA)/E_t$	$(EB)/E_t$
$(0,1,1,2,1)$	$(2,0,1,2,1)(A)$	$(2,4,0,3,1)(A)(B)$
$(0,1,1,3,1)$	$(2,0,1,3,1)(A)$	$(2,4,0,5,1)(A)(B)$
$(0,1,1,5,1)$	$(2,0,1,5,1)(A)$	
		$(3,1,0,2,1)(B)$
$(0,1,4,2,1)(A)$	$(2,0,4,2,1)(A)^2$	$(3,1,0,2,1)(B)$
$(0,1,4,5,1)(A)$	$(2,0,4,2,1)(A)^2$	$(3,1,0,5,1)(B)$
$(0,4,1,3,1)(B)$	$(3,0,4,2,1)(A)(B)$	$(3,4,0,3,1)(B)^2$
$(0,4,1,5.,1)(B)$	$(3,0,4,5,1)(A)(B)$	$(3,4,0,5,1)(B)^2$
$(0,4,4,5,1)(A)(B)$		

$(EAB)/E_t$	$(EQ)/E_t$
$(2,4,1,0,1)$	$(2,4,1,5,0)(A)(B)$
$(3,1,4,0,1)$	$(3,1,4,5,0)(A)(B)$
$(2,4,4,0,1)(A)^2(B)$	$(2,4,4,5,0)(A)^2(B)$
$(3,4,4,0,1)(A)(B)^2$	$(3,4,4,5,0)(A)(B)^2$

Fig. 11.2. Enzyme distribution expressions for the reaction sequence in Fig. 11.1.

The enzyme distribution expressions for the reaction sequence are given in Figure 11.2. Under the conditions specified, the rate of the enzymic reaction is

$$v = \left[k_{51} \frac{(EQ)}{E_t} \right] E_t \tag{11.1}$$

The rate equation in coefficient form is as follows:

$$v = \frac{\left[\begin{array}{l} (2,4,1,5,1)(A)(B) \\ (3,1,4,5,1)(A)(B) + (3,4,4,5,1)(A)(B)^2 + (2,4,4,5,1)(A)^2(B) \end{array} \right] E_t}{\begin{array}{llll} (0,1,1,2,1) & (2,0,1,2,1)(A) & (2,4,0,3,1)(A)(B) & (2,0,4,2,1)(A)^2 \\ (0,1,1,3,1) & (2,0,1,3,1)(A) & (2,4,0,5,1)(A)(B) & (2,0,4,5,1)(A)^2 \\ (0,1,1,5,1) & (2,0,1,5,1)(A) & (3,0,4,2,1)(A)(B) & \\ (3,1,0,2,1)(B) & (0,1,4,2,1)(A) & (3,0,4,5,1)(A)(B) & (2,4,4,5,0)(A)^2(B) \\ (3,1,0,3,1)(B) & (0,1,4,5,1)(A) & (0,4,4,5,1)(A)(B) & (2,4,4,0,1)(A)^2(B) \\ (3,1,0,5,1)(B) & (2,4,1,5,0)(A)(B) & (3,4,4,5,0)(A)(B)^2 & \\ (0,4,1,3,1)(B) & (3,1,4,5,0)(A)(B) & (3,4,4,0,1)(A)(B)^2 & \\ (0,4,1,5,1)(B) & (2,4,1,0,1)(A)(B) & & \\ (3,4,0,3,1)(B)^2 & (3,1,4,0,1)(A)(B) & & \\ (3,4,0,5,1)(B)^2 & & & \end{array}} \tag{11.2}$$

The first three numerator terms in eq. (11.2) contain (A) to the first power, while the last numerator term contains $(A)^2$. The terms in the first column of the denominator of eq. (11.2) consist of terms which contain (A) to the zero

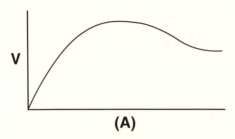

Fig. 11.3. Substrate-saturation curve described by a 2:2 rational polynomial when the enzyme is subject to substrate inhibition.

power, while the second and third columns of the denominator contain (A) to the first power and the last denominator column consists of terms which contain $(A)^2$. Thus, if A were the variable substrate, eq. (11.2) could be restated as

$$v = \frac{\alpha_1(A) + \alpha_2(A)^2}{\beta_0 + \beta_1(A) + \beta_2(A)^2} \tag{11.3}$$

Equation (11.3) is a 2:2 rational polynomial. When the general form of eq. (11.3) is applied to the reaction sequence shown in Figure 11.1, α_1 consists of two terms which contain (B) to the first power and one term which contains $(B)^2$. Furthermore, β_0 consists of terms which contain the concentration of the non-varied substrate to the zero, first and second powers. The latter is also true of β_1. The denominator coefficient β_2 contains (B) to the zero and first power. Equation (11.3) can describe curves of a variety of shapes, including a hyperbolic curve. That is, eq. (11.3) can, in essence, degrade into a 1:1 rational polynomial. One should not look at an equation such as eqs. (11.2) or (11.3) and attempt to predict what the shape of the curve should be. The actual shape of the curve described by such equations can be determined only by an analysis of the data obtained in experiments. Procedures for the analysis of the experimental data will be discussed in the following chapter. However, with regard to the reaction sequence portrayed in Figure 11.1, if k_{34} were very much smaller than the remaining rate constants, saturation of the enzyme with substrate B would force the reaction to follow the path leading to the slow step. If such were the case, the enzyme would be subject to substrate inhibition by B. This is yet another basis for substrate inhibition. It differs from those discussed earlier in that the earlier examples were described by 1:2 rational polynomials. In this case, the equation is a 2:2 rational function, and the substrate-saturation curve would appear as in Figure. 11.3.

11.2 The kinetic behavior of allosteric enzymes

The model for general enzyme inhibition portrayed in Figure. 3.1 was developed with the assumption that inhibition was total. That is, it was assumed that if the inhibitor combined with the enzyme to form an EI complex, the substrate could not combine with the EI complex to form an EAI complex. Further, it was assumed that when the inhibitor combined with the EA complex to form the EAI complex, this complex could not undergo the catalytic reaction to form the product. At this point, a model will be considered where inhibition is *not* total. However, the model will be modified to accommodate the interaction of the enzyme with an activator as well as an inhibitor. The letter M will identify the modifier regardless of whether M is a positive or negative modifier.

The concentration of the product in the reaction sequence shown in Figure 11.4 is assumed to be equal to zero so that the reaction is irreversible. The rate of the reaction is

$$v = \left[k_{31} \frac{(EP)}{E_t} + k_{64} \frac{(MEP)}{E_t} \right] E_t \tag{11.4}$$

The connection matrix and Q matrix are

$$U = \begin{vmatrix} 0 & A & 0 & M & 0 & 0 \\ 1 & 0 & 1 & 0 & M & 0 \\ 1 & 0 & 0 & 0 & 0 & 0 \\ 1 & 0 & 0 & 0 & A & 0 \\ 0 & 1 & 0 & 1 & 0 & 1 \\ 0 & 0 & 0 & 1 & 0 & 0 \end{vmatrix}, \quad Q = \begin{vmatrix} 2 & 4 & 0 \\ 1 & 3 & 5 \\ 1 & 0 & 0 \\ 1 & 5 & 0 \\ 2 & 4 & 6 \\ 4 & 0 & 0 \end{vmatrix}$$

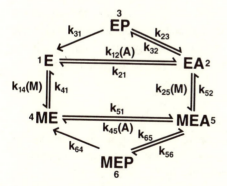

Fig. 11.4. A simple reaction sequence of an allosteric enzyme. Reactant M is an allosteric modifier which may be either an activator or an inhibitor.

The rate equation in coefficient form will not be presented here, but it can be anticipated that the general form of the equation will be identical to that of eq. (11.3). The coefficients in the equation will be zero, first or second order functions of the concentration of the modifier. The substrate-saturation curve might exhibit any of the shapes allowable in a 2:2 rational polynomial. The reaction sequence portrayed in Figure 11.4 is a rather good preliminary model of an allosteric enzyme. In many cases, these enzymes exhibit a sigmoidal substrate-saturation curve as this is one of the shapes which can be described by a 2:2 rational function.

11.3 Multiple enzymes catalyzing the conversion of a single substrate

A final example of an enzyme model which gives rise to a rate equation which is a 2:2 rational polynomial is exemplified by a mixture of two isozymes catalyzing the conversion of the same substrate at the same time. The rate equation for such a situation is given by eq. (11.5).

$$v = \frac{V_1(A)}{K_1 + (A)} + \frac{V_2(A)}{K_2 + (A)} \tag{11.5}$$

In eq. (11.5), V_1 and V_2 are the maximal velocities of isozymes 1 and 2, respectively, and K_1 and K_2 are their Michaelis constants. Equation (11.5) can be rearranged as follows

$$v = \frac{(V_1 K_2 + V_2 K_1)(A) + (V_1 + V_2)(A)^2}{K_1 K_2 + (K_1 + K_2)(A) + (A)^2}$$

Once again, the rate equation for this model is that of a 2:2 rational polynomial. The models presented here are three of many models which might be presented as examples of enzyme reaction sequences which give rise to rate equations which are higher order rational polynomials and which, for this reason may describe non-hyperbolic substrate-saturation curves.

Problems for Chapter 11

11.1 What would be the appearance of the equation to describe the relationship between the velocity and the concentration of the effector M for the reaction mechanism shown in Figure 11.4?

11.2 Derive the rate equation for the conversion of a single substrate to a product if the reaction were catalyzed by a mixture of three isozymes.

References

1. Koshland, D. E. (1958). Application of a theory of enzyme specificity to protein synthesis. *Proc. Natl. Acad. Sci. USA* **44**: 98–104.
2. Cross, D. G. and Fisher, H. F. (1970). The mechanism of glutamate dehydrogenase reaction. III The binding of ligands at multiple subsites and resulting kinetic effects. *J. Biol. Chem.* **245**: 2612–21.

12

Analysis of non-hyperbolic enzyme kinetics

The rate equation for an enzyme-catalyzed reaction derived in chapter 1 was a 1:1 rational polynomial. In chapter 3, the equation derived for substrate inhibition was a 1:2 rational polynomial, and in chapter 11 a 2:2 rational polynomial was derived for three separate enzyme models. The term "non-Michaelian" is sometimes applied to enzymes whose steady state kinetic behavior is described by rational functions of a order higher than 1:1. This is a misnomer. There was nothing "non-Michaelian" about the manner in which the equations were derived for the enzyme models considered in chapter 11. The fact is, that the general rate equation for an enzyme-catalyzed reaction is the following:

$$v = \frac{\sum\limits_{i=1}^{m} \alpha_i (A)^i}{\sum\limits_{i=0}^{n} \beta_i (A)^i}, \quad m \leq n, \alpha_i \geq 0, \beta_i \geq 0, \quad \text{where all } \beta_i \neq 0. \tag{12.1}$$

The stipulation $m \leq n$ is required to provide for saturation. Equation (12.1) can describe a variety of shapes which can vary from that of a rectangular hyperbola to a sigmoidal curve with multiple turning points. Thus, the substrate-saturation curve for an enzymic reaction may be hyperbolic or non-hyperbolic. The actual shape of the substrate-saturation curve is determined by an analysis of the data obtained from experiments. The purpose of the discussion in this chapter is to examine the restrictions which the various shapes of the substrate-saturation curve place on the parameters of eq. (12.1). Because the rate equations of most enzyme-catalyzed reactions which exhibit non-hyperbolic kinetic behavior can be defined by a 2:2 rational polynomial, emphasis will be placed on eq. (12.2).

$$v = \frac{\alpha_1 (A) + \alpha_2 (A)^2}{\beta_0 + \beta_1 (A) + \beta_2 (A)^2} \tag{12.2}$$

12.1 The use of analytical geometry to analyze a substrate-saturation curve

The basis for the analysis to be followed in this chapter was proposed by a very good enzyme kineticist, Jean Botts[1], and her treatment has been expanded upon by the publications of Bardsley and Childs[2-6]. The procedure which will be employed involves the basic principles of analytical geometry. There are two points on the curve which define the relationship between an independent variable and its dependent variable which can usually be identified. These points are where the value of the independent variable approaches zero and where it approaches infinity. One can determine the value of the dependent variable as these points are approached. Additional information can be obtained from the first derivative of the function, for this will provide information concerning the slope of the curve as these points are approached, but it will also provide information about any turning points that might occur between these points. Finally, the second derivative of the function provides information concerning the curvature of the curve as these points are approached. The second derivative will provide information about linear portions of the curve and about any inflection points which lie between these two points. This type of analysis will be performed on the rate equation and on the Lineweaver-Burk plot and one other linear form of the equation.

12.2 Analysis of the rate equation which is a 2:2 rational function

Reference to eq. (12.2) shows that the limiting values of v are $v_{\lim(A)\to 0} = 0$, $v_{\lim(A)\to\infty} = \alpha_2/\beta_2$. Thus the substrate-saturation curve passes through the origin and approaches a final value as (A) approaches infinity. It is significant that this final value is not necessarily a maximum value. Indeed, the final value could be equal to zero if $\alpha_2 = 0$ as is the case where the rate equation is a 1:2 rational function. The expression for the first derivative of eq. (12.2) is given by eq. (12.3).

$$\frac{dv}{d(A)} = \frac{\dfrac{dN}{d(A)}D - N\dfrac{dD}{d(A)}}{D^2} \tag{12.3}$$

where D is the denominator of eq. (12.2) and N is the numerator of eq. (12.2). Thus,

$$N = \alpha_1(A) + \alpha_2(A)^2 \tag{12.4}$$

$$\frac{dN}{d(A)} = \alpha_1 + 2\alpha_2(A) \tag{12.5}$$

$$\frac{d^2 N}{d(A)^2} = 2\alpha_2 \tag{12.6}$$

$$D = \beta_0 + \beta_1(A) + \beta_2(A)^2 \tag{12.7}$$

$$\frac{dD}{d(A)} = \beta_1 + 2\beta_2(A) \tag{12.8}$$

$$\frac{d^2 D}{d(A)^2} = 2\beta_2 \tag{12.9}$$

Substitution of eqs. (12.4), (12.5), (12.7) and (12.8) into eq. (12.3) gives

$$\frac{dv}{d(A)} = \frac{\alpha_1 \beta_0 + 2\alpha_2 \beta_0(A) + (\alpha_2 \beta_1 - \alpha_1 \beta_2)(A)^2}{(\beta_0 + \beta_1(A) + \beta_2(A)^2)^2} \tag{12.10}$$

It is immediately apparent from eq. (12.10) that the slope of the substrate-saturation curve will always be positive at low concentrations of the variable substrate. However, at high concentrations of substrate A, the slope may be positive or negative. The condition that must exist if the slope is to be negative at high concentrations of A is $\alpha_1\beta_2 \succ \alpha_2\beta_1$. This condition must exist if substrate inhibition is observed, and it is obvious that this condition is always met if the rate equation is a 1:2 rational function. As pointed out in chapter 11, a reaction sequence which gives rise to a rate equation which is a 2:2 rational polynomial may also exhibit substrate inhibition. The limiting values of the first derivative of eq. (12.2) are $dv/d(A)_{\lim(A)\to 0} = \alpha_1/\beta_0$, $dv/d(A)_{\lim(A)\to\infty} = 0$. Thus, α_1/β_0 is an apparent first order rate constant and this is the definition given to this apparent first order rate constant in chapter 1. If a sufficiently high concentration of the variable substrate can be attained, the substrate-saturation curve will approach a straight line asymptotically, and this line will be parallel to the (A) axis. This will be true regardless of whether the final portion of the substrate-saturation curve is a maximum value.

The expression for the second derivative of a rational polynomial is,

$$\frac{d^2 v}{d(A)^2} = \frac{d\left(\frac{d^2 N}{d(A)^2}D - \frac{d^2 D}{d(A)^2}N\right) - 2\frac{dD}{d(A)}\left(\frac{dN}{d(A)}D - \frac{dD}{d(A)}N\right)}{D^3} \tag{12.11}$$

Substitution of eqs. (12.4) through (12.9) into eq. (12.11) gives,

$$\frac{d^2v}{d(A)^2} = \frac{2\left[\begin{array}{c}\beta_0(\alpha_2\beta_0 - \alpha_1\beta_1) - 3\alpha_1\beta_0\beta_2(A) - 3\alpha_2\beta_0\beta_2(A)^2 \\ + \beta_2(\alpha_1\beta_2 - \alpha_2\beta_1)(A)^3\end{array}\right]}{(\beta_0 + \beta_1(A) + \beta_2(A)^2)^3} \qquad (12.12)$$

The limiting values of the second derivative of eq. (12.3) are

$$\frac{d^2v}{d(A)^2}_{\lim(A)\to 0} = \frac{2(\alpha_2\beta_0 - \alpha_1\beta_1)}{\beta_0^2}, \qquad \frac{d^2v}{d(A)^2}_{\lim(A)\to\infty} = 0.$$

From these limiting values, it is obvious that the initial portion of the curve may be concave upward or concave downward. At intermediate concentrations of the variable substrate, the curve is concave downward. The condition which must be met if the initial portion of the curve is to be concave upward is $\alpha_2\beta_0 > \alpha_1\beta_1$, and therefore, this is the condition that is met if the substrate-saturation curve is sigmoidal. The second derivative approaches a value of zero in an asymptotic manner as the concentration of substrate A approaches infinity. It should be recalled that the second derivative of a curve is equal to zero under two conditions. These conditions are when the curve passes through an inflection point and when the relationship is linear. Before the substrate-saturation curve becomes linear and parallel to the (A) axis at high concentrations of A, the second derivative may be positive or negative. The condition which must be met if it is to be positive in this region is $\alpha_1\beta_2 > \alpha_2\beta_1$. This is the condition necessary for substrate inhibition. Since the substrate-saturation curve is always concave downward at intermediate concentrations of the variable substrate, the curve passes through an inflection point at relatively low concentrations of A if the curve is sigmoidal, and it will pass through an inflection point at relatively high concentrations of A if the enzyme is subject to substrate inhibition.

12.3 Analysis of the Lineweaver-Burk plot of a 2:2 rational polynomial

The Lineweaver-Burk equation of an enzymic reaction whose rate equation is a 2:2 rational polynomial is given in eq. (12.13).

$$\frac{1}{v_a} = \frac{\beta_2 + \beta_1\frac{1}{(A)} + \beta_0\left(\frac{1}{(A)}\right)^2}{\alpha_2 + \alpha_1\frac{1}{(A)}} \qquad (12.13)$$

Equation (12.13) would appear to describe a nonlinear plot, but it will be noted later in this discussion that conditions can be met in which the plot is linear, that is, conditions under which the substrate-saturation curve is hyperbolic. The limiting values of eq. (12.13) are $1/v_{\lim 1/(A) \to 0} = \beta_2/\alpha_2$, $1/v_{\lim 1/(A) \to \infty} = \infty$. The following relationships are used to obtain the equations for the first and second derivatives of eq. (12.13).

$$N = \beta_2 + \beta_1 \frac{1}{(A)} + \beta_0 \left(\frac{1}{(A)}\right)^2 \tag{12.14}$$

$$\frac{dN}{d(1/(A))} = \beta_1 + 2\beta_0 \frac{1}{(A)} \tag{12.15}$$

$$\frac{d^2N}{d(1/(A))^2} = 2\beta_0 \tag{12.16}$$

$$D = \alpha_2 + \alpha_1 \frac{1}{(A)} \tag{12.17}$$

$$\frac{dD}{d(1/(A))} = \alpha_1 \tag{12.18}$$

$$\frac{d^2D}{d(1/(A))^2} = 0 \tag{12.19}$$

The expression for the first derivative of eq. (12.13) is obtained by substituting eqs. (12.14), (12.15), (12.17) and (12.18) into eq. (12.3).

$$\frac{d(1/v)}{d(1/(A))} = \frac{(\alpha_2\beta_1 - \alpha_1\beta_2) + 2\alpha_2\beta_0 \frac{1}{(A)} + \alpha_1\beta_0 \left(\frac{1}{(A)}\right)^2}{\left(\alpha + \alpha_1 \frac{1}{(A)}\right)^2} \tag{12.20}$$

The limiting values of the first derivative of the Lineweaver-Burk equation are

$$\frac{d(1/v)}{d(1/(A))_{\lim 1/(A) \to 0}} = \frac{\alpha_2\beta_1 - \alpha_1\beta_2}{\alpha_2^2}, \quad \frac{d(1/v)}{d(1/(A))_{\lim 1/(A) \to \infty}} = \frac{\beta_0}{\alpha_1}$$

The initial slope of the Lineweaver-Burk plot may be positive or negative. The condition which must be satisfied if the initial slope is to be negative is $\alpha_1\beta_2 \succ \alpha_2\beta_1$, and it has been established previously that this is the condition required for substrate inhibition. At higher values of $1/(A)$, the Lineweaver-Burk equation becomes linear and the slope of the linear portion is β_0/α_1.

The expression for the second derivative of the Lineweaver-Burk equation is obtained by substitution of eqs. (12.14) through (12.19) into eq. (12.11).

$$\frac{d^2(1/v)}{d(1/(A))^2} = \frac{2(\alpha_1^2 \beta_2 - \alpha_1 \alpha_2 \beta_1 + \alpha_2^2 \beta_0)}{\left(\alpha_2 + \alpha_1 \dfrac{1}{(A)}\right)^3} \qquad (12.21)$$

The limiting values of the second derivative are

$$\frac{d^2(1/v)}{d(1/(A))^2_{\lim 1/(A) \to 0}} = \frac{2[\alpha_1^2 \beta_2 - \alpha_1 \alpha_2 \beta_1 + \alpha_2^2 \beta_0]}{\alpha_2^3}$$

$$\frac{d^2(1/v)}{d(1/(A))^2_{\lim 1/(A) \to \infty}} = 0$$

The Lineweaver-Burk plot becomes linear at high values of $1/(A)$ as indicated by the fact that the second derivative is equal to zero in that region. The striking feature about eq. (12.21) is that while the numerator contains both positive and negative terms, the independent variable does not appear in any of the numerator terms. Therefore, there are no inflection points in the Lineweaver-Burk plot of a 2:2 rational function. However, at lower values of $1/(A)$, the second derivative can be positive or negative or equal to zero. The second derivative will be equal to zero in this region if $\beta_1 = (\alpha_2/\alpha_1)\beta_0 + (\alpha_1/\alpha_2)\beta_2$. If this condition is satisfied, the Lineweaver-Burk plot is linear, and the substrate-saturation curve is hyperbolic. If such is the case, in essence, the equation degrades into a 1:1 rational polynomial. Figure 12.1a shows the substrate-saturation curve and Figure 12.1b shows the Lineweaver-Burk plot for this situation.

There are conditions under which the second derivative of the Lineweaver-Burk plot will be positive and the curve will, therefore, be

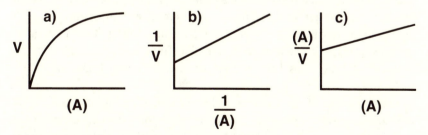

Fig. 12.1. Plots which characterize an enzyme-catalyzed reaction described by a hyperbolic substrate-saturation curve (1:1 rational polynomial). a) Substrate-saturation curve. b) Lineweaver-Burk plot. c) Hanes plot.

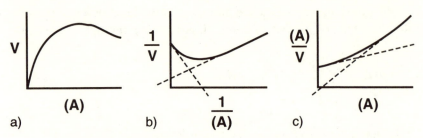

Fig. 12.2. Plots which characterize a 2:2 rational polynomial with substrate inhibition. a) Substrate-saturation curve. b) Lineweaver-Burk plot. c) Hanes plot. The broken lines are extensions of the tangent to the plots at the *y* axis and extensions of the linear asymptote.

concave upward prior to the linear segment at high values of $1/(A)$. One of these conditions can be visualized easily by re-arranging the numerator terms of eq. (12.21) as $\alpha_1(\alpha_1\beta_2 - \alpha_2\beta_1) + \alpha_2^2\beta_0$. The difference between the first two terms in the foregoing expression will be positive if $\alpha_1\beta_2 \succ \alpha_2\beta_1$ and this is the condition for substrate inhibition. The third term will always be positive, and the Lineweaver-Burk plot will be concave upward in the case of substrate inhibition until it finally becomes linear. This conclusion has been reached previously, but the foregoing is a quantitative analysis of the relative magnitude of the parameters of a 2:2 rational polynomial which will give rise to this phenomenon. Figures 12.2a and 12.2b portray the substrate-saturation curve and the Lineweaver-Burk plot for substrate inhibition for a 2:2 rational function. Another set of conditions which will result in the nonlinear region of the Lineweaver-Burk plot having a positive second derivative can be visualized by rearranging the numerator terms of eq. (12.21) as $\alpha_1^2\beta_2 + \alpha_2(\alpha_2\beta_0 - \alpha_1\beta_1)$. This expression will be positive if $\alpha_2\beta_0 \succ \alpha_1\beta_1$, and this is the condition necessary if the substrate-saturation curve is to be sigmoidal. This may arise if the enzyme contains an allosteric site, or if the enzyme is polymeric and has multiple active subunits which interact cooperatively. The kinetic behavior of enzymes with multiple active subunits will be discussed in the following chapter. The appearance of the substrate-saturation curve and the Lineweaver-Burk plot for an enzyme exhibiting this behavior are shown in Figures 12.3a and 12.3b, respectively.

The numerator of eq. (12.21) provides one further possibility: namely, the second derivative of the curve prior to the linear portion could be negative if $\beta_1 \succ (\alpha_1/\alpha_2)\beta_2 + (\alpha_2/\alpha_1)\beta_0$. Thus, the Lineweaver-Burk plot would be concave downward initially. This behavior is characteristic of enzymes which exhibit substrate activation or it may be exhibited by isozymes competing

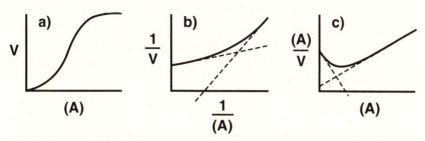

Fig. 12.3. Plots which characterize a 2:2 rational polynomial when the substrate-saturation curve is sigmoidal.

for the same substrate, or it is exhibited by polymeric enzymes which are subject to negative cooperativity. In the case of substrate activation, the apparent maximal velocity increases as the substrate concentration increases. Figures 12.4a and 12.4b show the substrate-saturation curve and the Lineweaver-Burk plot for this type of kinetic behavior. At intermediate concentrations of the variable substrate, there is a rather wide region where the velocity is close to a linear function of (A).

In Figures. 12.2, 12.3 and 12.4, lines have been drawn tangent to the Lineweaver-Burk plots at the point where the Lineweaver-Burk plot intersects the $1/v$ axis and also to the asymptote which is approached at high concentrations of the variable substrate. It is easy to obtain the equation for the line tangent to the curve at the intersection of the $1/v$ axis. The slope of the line is the first derivative of the Lineweaver- Burk plot when $1/(A) = 0$. The intercept of the line is the point at which the Lineweaver-Burk plot intersects the $1/v$ axis. The equation of the line is,

$$Y_T = \left(\frac{\alpha_2 \beta_1 - \alpha_1 \beta_2}{\alpha_2^2} \right) \frac{1}{(A)} + \frac{\beta_2}{\alpha_2} \tag{12.22}$$

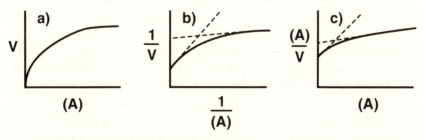

Fig. 12.4. Plots of a 2:2 rational polynomial which are characteristic of substrate activation, or negative cooperativity, or multiple enzymes catalyzing the conversion of a common substrate.

The slope of the line tangent to the asymptote is the limiting value of the first derivative of the Lineweaver-Burk plot as $1/(A)$ approaches infinity, but the intercept of this line is not immediately available. Fortunately, this intercept can be obtained by a little simple mathematics. The point at which any curve intersects the y axis is

$$Y_0 = Y_i - X_i \frac{dY}{dX} \tag{12.23}$$

where Y_0 is the point of intersection, Y_i is the function at any point and, in the present case, Y_i can be replaced by eq. (12.13). The term X_i can be replaced by $1/(A)$ and dY/dX can be replaced by eq. (12.20).

$$\left(\frac{1}{v}\right)_0 = \frac{\alpha_2\beta_2 + 2\alpha_1\beta_2\dfrac{1}{(A)} + (\alpha_1\beta_1 - \alpha_2\beta_0)\left(\dfrac{1}{(A)}\right)^2}{\left(\alpha_2 + \alpha_1\dfrac{1}{(A)}\right)^2} \tag{12.24}$$

Equation (12.24) is the point of intersection of the $1/v$ axis of a line drawn tangent at any point of the Lineweaver-Burk plot of a 2:2 rational function. When the value of $1/(A)$ approaches infinity, the only significant terms in eq. (12.24) are those which contain $(1/(A))^2$, and the equation of the asymptote is,

$$Y_A = \frac{\beta_0}{\alpha_1}\frac{1}{(A)} + \frac{\alpha_1\beta_1 - \alpha_2\beta_0}{\alpha_1^2} \tag{12.25}$$

Thus, if the asymptote intersects the $1/v$ axis below the $1/(A)$ axis, the substrate-saturation curve must be sigmoidal.

Inspection of Figures. 12.2, 12.3 and 12.4 reveals that the line drawn at a tangent to the Lineweaver-Burk plot at the intersection of the $1/v$ axis will intersect the asymptote at some point. The coordinates of this latter point of intersection can be obtained by setting eq. (12.22) equal to eq. (12.25). These coordinates are found to be $\alpha_2/\alpha_1, \beta_1/\alpha_1$. These parameters, together with those already defined, give estimates of four parameters, namely, $\beta_0/\alpha_1, \beta_2/\alpha_2, \alpha_2/\alpha_1, \beta_1/\alpha_1$. From the foregoing, it is possible to obtain estimates of ratios of any of the parameters. The reader may feel that these ratios do not provide a satisfactory analysis of the curves, but it should be recalled that, in the case of the original rate equation derived in chapter 1, $V_m = \alpha_1/\beta_1$, $K_m = \beta_0/\beta_1$. Hence, the 2:2 rational polynomial can be analyzed in a manner comparable to a 1:1 rational polynomial. There are statistical programs which provide for the analysis of higher order rational polynomials, but a discussion of these involves statistical procedures that go beyond the scope of this book.

12.4 Analysis of the Hanes plot of a 2:2 rational polynomial

An alternative to the Lineweaver-Burk plot is the plot of $(A)/v$ versus (A). For convenience, this will be called the Hanes plot because C. S. Hanes was one of the first to employ it[7]. Equation (12.26) is the primary equation for this plot.

$$\frac{(A)}{v} = \frac{\beta_0 + \beta_1(A) + \beta_2(A)^2}{\alpha_1 + \alpha_2(A)} \tag{12.26}$$

The limiting values are $(A)/v_{\lim(A)\to0} = \beta_0/\alpha_1$, $(A)/v_{\lim(A)\to\infty} = \infty$. The following relationships are useful in writing the first and second derivatives.

$$N = \beta_0 + \beta_1(A) + \beta_2(A)^2 \tag{12.27}$$

$$\frac{dN}{d(A)} = \beta_1 + 2\beta_2(A) \tag{12.28}$$

$$\frac{d^2N}{d(A)^2} = 2\beta_2 \tag{12.29}$$

$$D = \alpha_1 + \alpha_2(A) \tag{12.30}$$

$$\frac{dD}{d(A)} = \alpha_2 \tag{12.31}$$

$$\frac{d^2D}{d(A)^2} = 0 \tag{12.32}$$

The first derivative of the Hanes plot is

$$\frac{d\left(\frac{(A)}{v}\right)}{d(A)} = \frac{(\alpha_1\beta_1 - \alpha_2\beta_0) + 2\alpha_1\beta_2(A) + \alpha_2\beta_2(A)^2}{[\alpha_1 + \alpha_2(A)]^2} \tag{12.33}$$

The limiting values of the first derivative are

$$\frac{d\left(\frac{(A)}{v}\right)}{d(A)_{\lim(A)\to0}} = \frac{\alpha_1\beta_1 - \alpha_2\beta_0}{\alpha_1^2}, \quad \frac{d\left(\frac{(A)}{v}\right)}{d(A)_{\lim(A)\to\infty}} = \frac{\beta_2}{\alpha_2}$$

The initial slope of the Hanes plot may be positive or negative. It will be negative if the substrate-saturation curve is sigmoidal. This is shown in Fig. 12.3c, and this is probably the most sensitive indicator of a sigmoidal substrate-saturation curve. The second derivative of the Hanes plot is given by eq. (12.34).

$$\frac{d^2\left(\frac{(A)}{v}\right)}{d(A)^2} = \frac{2(\alpha_1^2\beta_2 - \alpha_1\alpha_2\beta_1 + \alpha_2^2\beta_0)}{[\alpha_1 + \alpha_2(A)]^3} \tag{12.34}$$

The numerator of the second derivatives of the Lineweaver-Burk and the Hanes equations, when the rate equation is a 2:2 rational polynomial, are identical. Therefore the previous discussion of the second derivative of the Lineweaver-Burk plot also applies to the Hanes plot.

It is well known that, when the rate equation is a 1:1 rational function, the Michaelis constant is equal to the concentration of the variable substrate which gives half maximal velocity. This was shown to be correct in chapter 1, and is also shown by the following:

$$\frac{\alpha_1}{2\beta_1} = \frac{\alpha_1(A)}{\beta_0 + \beta_1(A)}$$

$$(A) = \frac{\beta_0}{\beta_1}$$

The parameter β_0/β_1 is defined as K_m. It is shown easily that, when the rate equation is a 2:2 rational function, the substrate concentration which gives half of the limiting velocity as the substrate concentration approaches infinity is not a simple parameter which can be obtained in studies of the steady state behavior of the enzyme. This is shown as follows:

$$\frac{\alpha_2}{2\beta_2} = \frac{\alpha_1(A) + \alpha_2(A)^2}{\beta_0 + \beta_1(A) + \beta_2(A)^2}$$

$$(A) = \frac{\alpha_2\beta_1 - 2\alpha_1\beta_2 + \sqrt{(\alpha_2\beta_1 - 2\alpha_1\beta_2)^2 + 4\alpha_2^2\beta_0\beta_2}}{2\alpha_2\beta_2}$$

Thus, if the substrate-saturation curve is sigmoidal, for example, the substrate concentration which gives half maximal velocity is not equal to a simple parameter such as a Michaelis constant. The treatment given earlier in this chapter has shown how steady state parameters can be obtained if the rate equation is a 2:2 rational polynomial, but the substrate concentration which gives half maximal velocity does not provide an estimate of one of these parameters. This does not detract from the significance that the concentration of substrate which gives half maximal velocity might have in metabolic considerations.

12.5 The kinetic behavior of higher order rational polynomials

The rate equation may be a higher order than a 2:2 rational polynomial[4]. In most of these cases, the equation degrades, in essence, into a 2:2 rational function. For this reason, higher order functions will not be considered in detail here. The distinguishing feature between a 2:2 rational function and higher order functions is that a 2:2 function can exhibit only one turning point. In contrast, a rate equation which is a 3:3 rational function can exhibit two turning points. A rate equation which is a 4:4 rational function can exhibit three turning points. However, it must be borne in mind that in order to determine if a substrate-saturation curve has multiple turning points, it is necessary to employ high concentrations of the variable substrate. It is essential that the experiments be conducted under conditions of constant ionic strength. It is also essential to establish that the multiple turning points are real, and not merely the reflection of experimental variations.

12.6 Problems for chapter 12

12.1 The equation derived in chapter 11 for two enzymes catalyzing the conversion of a common substrate to a product was

$$v = \frac{(V_1 K_2 + V_2 K_1)(A) + (V_1 + V_2)(A)^2}{K_1 K_2 + (K_1 + K_2)(A) + (A)^2}$$

Using the procedures described in this chapter, determine whether this equation could describe a sigmoidal substrate-saturation curve.

12.2 Could the equation in problem 12.1 describe a hyperbolic substrate-saturation curve, and if so, what condition must exit?

12.3 Could the equation in problem 12.1 describe a substrate-saturation curve which would be characteristic of negative cooperativity or substrate activation?

12.4 Could the equation in problem 12.1 describe substrate inhibition?

References

1. Botts, J. (1958). Typical behavior of some models of enzyme action. *Trans. Faraday Soc.* **54**: 593–604.
2. Bardsley, W. G. and Childs, R. E. (1975). Sigmoidal curves, non-linear double-reciprocal plots and allosterism. *Biochem. J.* **149**: 313–28
3. Childs, R. E. and Bardsley, W. G. (1976). An analysis of non-linear Eadie-Hofstee-Scatchard representations of ligand-binding and initial rate for allosteric and other complex enzyme mechanism. *J. Theor. Biol.* **63**: 1–18.
4. Bardsley, W. G. (1977). The 3:3 function in enzyme kinetics. Possible shapes of v/S and $(1/v)(1/S)$ plots for third degree steady-state equations. *J. Theor. Biol.* **65**: 281–316.
5. Bardsley, W. G. and Waight, R. D. (1978). The determination of positive and negative co-operativity with allosteric enzymes and the interpretation of sigmoidal curves and non-linear double reciprocal plots for the MWC and KNF models. *J. Theor. Biol.* **70**: 135–56.
6. Bardsley, W. G., Leff, P., Kavanagh, J. and Waight, R. D. (1980). Deviations from Michaelis-Menten kinetics. *Biochem. J.* **187**: 739–65.
7. Hanes, C. S. (1932). The effect of starch concentration upon the velocity of hydrolysis by amylase of germinated barley. *Biochem. J.* **26**: 1406–21.

13

The effect of subunit interactions on enzyme kinetics

In addition to the enzyme models discussed in chapter 11, subunit interactions can also give rise to non-hyperbolic enzyme kinetics, and allosteric enzymes often are polymeric proteins in which the regulatory site is present on a separate regulatory subunit. The effect of subunit interactions on substrate binding in polymeric proteins in which more than one subunit contains an active site has been discussed in a number of classical publications[1,2]. Ricard and his colleagues have discussed the effect of subunit interactions on the steady state kinetic behavior of enzymes[3,4]. The treatment of Ricard, Mouttet and Nari will be presented here[3].

13.1 The effect of subunit interactions on rate constants

For the purpose of this analysis, the free energy of activation, $\Delta G^{\#}$, of any reaction step catalyzed by a polymeric enzyme is split into four components in the Ricard treatment. The free energy of activation is divided into the following components as a matter of convenience.

1 The intrinsic free energy of activation of trans-conformational change, $\Delta G_t^{\#*}$.
2 The intrinsic free energy of activation due to non-transconformational causes, $\Delta G_n^{\#*}$.
3 The contribution of subunit interactions to transconformational free energy of activation, $\sum \Delta G_t^{\#s}$.
4 The contribution of subunit interactions to non-transconformational free energy of activation, $\sum \Delta G_n^{\#s}$.

The symbolism employed here is slightly different from that employed by Ricard. The non-transconformational processes include ligand binding and

153

dissociation and also the catalytic steps. If the enzyme were monomeric, $\Delta G^\# = \Delta G_t^{\#*} + \Delta G_n^{\#*}$. On the other hand, for a completely rigid polymeric enzyme, $\Delta G^\# = \Delta G_n^{\#*} + \sum \Delta G_t^{\#s}$. If the direct neighborhood of the active site is not affected by subunit interactions, $\sum \Delta G_t^{\#s} = 0$. This is usually a valid assumption, and it allows for a less complicated insight into the effect of subunit interaction on the steady state kinetic behavior of the enzyme. Ricard and Noat[4-6] have given a more general treatment of the effect of subunit interactions which does not require this restriction.

For the purpose of introducing the basic concept, it will be assumed that two conformations are accessible to the protomers of the polymeric enzyme. To maintain consistency with the symbolism employed in this text, these conformations will be referred to as F and G. The total subunit interactions within the polymer, at any given time, can involve l FF interactions, m FG interactions and n GG interactions. The total contribution of subunit interaction to the free energy of activation is,

$$\sum \Delta G_t^{\#s} = l \Delta G_{FF}^{\#s} + m \Delta G_{FG}^{\#s} + n \Delta G_{GG}^{\#s} \tag{13.1}$$

The relationship of a rate constant to the free energy of activation is given by eq. (13.2).

$$k = \frac{k_B T}{h} e^{-\Delta G^\#/RT} \tag{13.2}$$

where k_B is the Boltzman constant, h is the Planck constant, T is absolute temperature and R is the gas constant. Substitution of the expressions for intrinsic free energy of activation and the sum of the free energies of activation for subunit interactions into eq. (13.2) gives

$$k = \frac{k_B T}{h} e^{-(\Delta G_n^{\#*} + \Delta G_t^{\#*})/RT} e^{-l(\Delta G_{FF}^{\#s})/RT} e^{-m(\Delta G_{FG}^{\#s})/RT} e^{-n(\Delta G_{GG}^{\#s})/RT}$$

It is convenient to separate the right-hand side of eq. (13.3) into four components. The first component is an intrinsic rate constant.

$$k^* = \frac{k_b T}{h} e^{-(\Delta G_t^{\#*} + \Delta G_n^{\#*})/RT} \tag{13.4}$$

The remaining components can be defined as follows

$$\alpha_{FF} = e^{-\Delta G_{FF}^{\#s}/RT} \tag{13.5}$$

$$\alpha_{FG} = e^{-\Delta G_{FG}^{\#s}/RT} \tag{13.6}$$

$$\alpha_{GG} = e^{-\Delta G_{GG}^{\#s}/RT} \tag{13.7}$$

Equation (13.4) can be expressed as

$$k = k^* \alpha_{FF}^l \alpha_{FG}^m \alpha_{GG}^n \qquad (13.8)$$

It is obvious that the α values cannot be equal to zero, and that when they have a value of unity there is no subunit interaction. These are called the coefficients of subunit interaction. The smaller the coefficient of subunit interaction, the smaller is the free energy of subunit interaction, and therefore the more probable will be the interaction between the subunits.

13.2 The effect of sequential subunit interactions

A dimeric enzyme will be considered throughout this discussion. This allows presentation of the concepts in a simple manner, but it does not provide the full range of types of subunit interactions that are available in the case of proteins consisting of three or more subunits. Koshland, *et al.* proposed a number of ways in which the subunits of a protein might interact[2]. These include a linear, a square and a tetrahedral interaction scheme. However, there is no distinction between these subunit interaction schemes in the case of a dimeric protein, and the exponents in eq. (13.8) are all equal to 1 in a dimeric protein.

In Figure 13.1 the left-hand sequence portrays the apparent, or phenomenological, process, while the right-hand sequence portrays the intrinsic process in which subunit interaction is not involved. The F conformation is depicted as a circle while the G conformation is depicted as a square. The sequential scheme assumes that the change in conformation takes place as the substrate binds to the subunit, and only the subunit to which the substrate is bound undergoes a conformational change. The apparent, or phenomenological, rate constant for ligand binding, k_1, is multiplied by 2 because

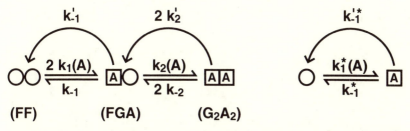

Fig. 13.1. Model of a dimeric enzyme in which the subunits interact sequentially. The addition of substrate converts the subunit to which it binds from the F conformation to the G conformation. The model on the right of the figure represents the intrinsic process.

the substrate could bind to either subunit. Likewise, k_2 and k'_2 are multiplied by 2 because the substrate could dissociate or the product could be formed and be dissociated from either subunit. The parameters obtained from studies of the steady state behavior of the polymeric enzyme are apparent parameters. It is desirable to express these in terms of the intrinsic parameters, and this is accomplished through the relationship given in eq. (13.8).

An apparent Michaelis constant can be written for the interaction of substrate with each subunit.

$$K_1^{app} = \frac{k_{-1} + k'_1}{2k_1} \tag{13.9}$$

$$K_2^{app} = \frac{2(k_{-2} + k'_2)}{k_2} \tag{13.10}$$

The intrinsic Michaelis constant applies to the right-hand sequence in Figure 13.1.

$$K^* = \frac{k^*_{-1} + k'^*_1}{k^*_1} \tag{13.11}$$

By reference to Figure 13.1 it is seen that in the absence of the substrate, the free dimer contains only F protomers, and that, in the case of a dimeric enzyme, there is only one FF interaction. The FGA species allows one FG interaction and the G_2A_2 species allows one GG interaction. Therefore, the observed rate constants are

$$k_1 = \alpha_{FF} k^*_1 \tag{13.12}$$

$$k_{-1} = \alpha_{FG} k^*_{-1} \tag{13.13}$$

$$k'_1 = \alpha_{FG} k'^*_1 \tag{13.14}$$

$$k_2 = \alpha_{FG} k^*_1 \tag{13.15}$$

$$k_{-2} = \alpha_{GG} k^*_{-1} \tag{13.16}$$

$$k'_2 = \alpha_{GG} k'^*_1 \tag{13.17}$$

Substitution of eqs. (13.12)–(13.14) into eq. (13.9) gives

$$K_1^{app} = \frac{\alpha_{FG}}{2\alpha_{FF}} K^* \tag{13.18}$$

Likewise, substitution of eqs. (13.15)–(13.17) into eq. (13.10) gives

$$K_2^{app} = \frac{2\alpha_{GG}}{\alpha_{FG}} K^* \tag{13.19}$$

The following relationships are recognized easily.

$$E_t = (FF) + (FGA) + (G_2A_2) \tag{13.20}$$

$$v = \alpha_{FG} k_1'^* (FGA) + 2\alpha_{GG} k_2'^* (G_2A_2) \tag{13.21}$$

$$(FGA) = \frac{(A)}{K_1^{app}} (FF) = \frac{2\alpha_{FF}(A)}{\alpha_{FG} K^*} (FF) \tag{13.22}$$

$$(G_2A_2) = \frac{(A)}{K_2^{app}} (FGA) = \frac{\alpha_{FF}(A)^2}{\alpha_{GG} K^{*2}} (FF) \tag{13.23}$$

$$(FF) = \frac{K^{*2} E_t}{K^{*2} + \dfrac{2\alpha_{FF}}{\alpha_{FG}} K^*(A) + \dfrac{\alpha_{FG}}{\alpha_{GG}} (A)^2} \tag{13.24}$$

The velocity of the reaction is

$$v = \frac{2\alpha_{FF} k_1'^* E_t(A)[K^* + (A)]}{K^{*2} + \dfrac{2\alpha_{FF}}{\alpha_{FG}} K^*(A) + \dfrac{\alpha_{FG}}{\alpha_{GG}} (A)^2} \tag{13.25}$$

Equation (13.25) is the identical form of eq. (11.3) with the following relationships between the parameters.

$$\alpha_1 = 2\alpha_{FF} k_1'^* K^* E_t, \quad \alpha_2 = 2\alpha_{FF} k_1'^* E_T,$$

$$\beta_0 = K^{*2}, \quad \beta_1 = \frac{2\alpha_{FF}}{\alpha_{FG}} K^*, \quad \beta_2 = \frac{\alpha_{FF}}{\alpha_{GG}}$$

If $\alpha_{FF} = \alpha_{FG} = \alpha_{GG}$, the velocity of the reaction becomes a 1:1 rational polynomial. That is, the two subunits are completely independent of one another and they behave as two separate enzymes with identical Michaelis constants and maximal velocities.

$$v = \frac{2\alpha_{FF} k_1'^* E_t(A)}{K^* + (A)} \tag{13.26}$$

It was shown in the previous chapter that the condition which must be satisfied if the rate equation is to describe a sigmoidal curve is $\alpha_2 \beta_0 > \alpha_1 \beta_1$. If the values from eq. (13.25) are substituted into this expression, the requirement for a sigmoidal curve is found to be $1/2 > \alpha_{FF}/\alpha_{FG}$. Thus the coefficient of subunit interaction of the FG subunits must be greater than twice that of the FF subunit interaction if positive cooperativity is to be observed. Since the FF interaction is more probable than the FG interaction, it is necessary to drive the reaction with increased substrate concentration. However, once the FGA species is formed, it readily decomposes to give rise to the product and the more probable FF species.

The necessary condition, if substrate inhibition is to be observed, is $\alpha_1 \beta_2 > \alpha_2 \beta_1$. The substitution of values from eq. (13.25) into the foregoing expression gives $(1/2) > (\alpha_{gg}/\alpha_{fg})$. Thus if substrate inhibition is to be observed, the GG subunit interaction must be more than twice as probable as the FG interaction. If such is the case, the $G_2 A_2$ species serves as a partial "sink" for the enzyme. Finally, in chapter 12 it was concluded that the substrate-saturation curve would be hyperbolic if $\beta_1 = (\alpha_2/\alpha_1)\beta_0 + (\alpha_1/\alpha_2)\beta_2$, and that negative cooperativity would be observed if β_1 were greater than the right-hand side of the expression. Therefore, the curve will be hyperbolic if $2 = (\alpha_{FG}/\alpha_{FF}) + (\alpha_{FG}/\alpha_{GG})$. This will be observed if $\alpha_{FF} = \alpha_{FG} = \alpha_{GG}$, but this conclusion was earlier reached intuitively. If 2 is greater than the right-hand side of the expression, eq. (13.25) describes negative cooperativity and this is promoted if the FG interaction is more probable than the FF or GG interactions.

It is significant that, in the case of the sequential conformational transitions, the shape of the substrate-saturation curve is determined by the coefficients of subunit interaction. It will be seen that this is unique to the mechanism of sequential conformational transitions.

13.3 The effect of partially concerted subunit interactions

A partially concerted model of subunit interaction is portrayed in Figure 13.2. In addition to the F and G conformations, this model contains a K conformation. The binding of substrate to a subunit in the F conformation is associated with the conversion of the subunit to the G conformation, and the binding to the active site of one subunit causes the transition of the remaining subunit from the F to the K conformation. The binding of a molecule of substrate to the subunit in the K conformation is associated with the transition of the subunit to the G conformation. Thus, there are two intrinsic processes. One intrinsic process is the binding of substrate to the active site of one subunit in the F conformation and the transition of that subunit to the G conformation. The second intrinsic process is the binding of substrate to the active site of a subunit in the K conformation and the transition of that subunit to the G conformation. The two apparent

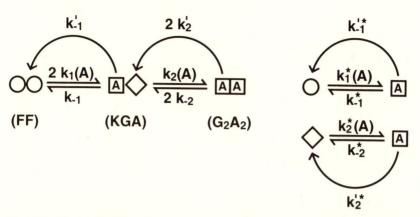

Fig. 13.2. Model of a dimeric enzyme in which the subunits interact in a partially concerted manner. The two models on the right of the figure represent the intrinsic processes.

Michaelis constants for the sequence in Figure 13.2 are

$$K_1^{app} = \frac{k_{-1} + k_1'}{2k_1} \tag{13.27}$$

$$K_2^{app} = \frac{2(k_{-2} + k_2')}{k_2} \tag{13.28}$$

The intrinsic Michaelis constants are

$$K_1^* = \frac{k_{-1}^* + k_1'^*}{k_1^*} \tag{13.29}$$

$$K_2^* = \frac{k_{-2}^* + k_2'^*}{k_2^*} \tag{13.30}$$

The observed rate constants are related to the intrinsic as follows

$$k_1 = \alpha_{FF} k_1^* \tag{13.31}$$

$$k_{-1} = \alpha_{KG} k_{-1}^* \tag{13.32}$$

$$k_1' = \alpha_{KG} k_1'^* \tag{13.33}$$

$$k_2 = \alpha_{KG} k_2^* \tag{13.34}$$

$$k_{-2} = \alpha_{GG} k_2^* \tag{13.35}$$

$$k_2' = \alpha_{GG} k_2'^* \tag{13.36}$$

Thus,

$$K_1 = \frac{\alpha_{KG}}{2\alpha_{FF}} K_1^* \tag{13.37}$$

$$K_2 = \frac{2\alpha_{GG}}{\alpha_{KG}} K_2^* \tag{13.38}$$

The following expressions are obtained from the previous relations.

$$E_t = (FF) + (KGA) + (G_2A_2) \tag{13.39}$$

$$(KGA) = \frac{2\alpha_{FF}(A)}{\alpha_{KG} K_1^*}(FF) \tag{13.40}$$

$$(G_2A_2) = \frac{\alpha_{FF}(A)^2}{\alpha_{GG} K_1^* K_2^*}(FF) \tag{13.41}$$

$$(FF) = \frac{K_1^* K_2^* E_t}{K_1^* K_2^* + \dfrac{2\alpha_{FF} K_2^*}{\alpha_{KG}}(A) + \dfrac{\alpha_{FF}}{\alpha_{GG}}(A)^2} \tag{13.42}$$

The rate of the reaction is

$$v = \frac{2\alpha_{FF} E_t(A)[k_1'^* K_2^* + k_2'^*(A)]}{K_1^* K_2^* + \dfrac{2\alpha_{FF} K_2^*}{\alpha_{KG}}(A) + \dfrac{\alpha_{FF}}{\alpha_{GG}}(A)^2} \tag{13.43}$$

An analysis of the 2:2 rational polynomial expressed in eq. (13.43) shows that the equation can describe all the shapes possible in the case of sequential subunit interactions. In this model of partially concerted subunit interaction, the intrinsic rate constants, the intrinsic Michaelis constants and the coefficients of subunit interaction contribute to the parameters of the rate equation. For example, if the substrate-saturation curve is to be sigmoidal, indicating positive cooperativity, $(1/2) > (\alpha_{FF}/\alpha_{KG})\cdot(k_1'^*/k_2'^*)(K_2^*/K_1^*)$. This is in contrast to the sequential subunit interaction model. While these two models differ in this regard, it is important to realize that steady state kinetic studies do *not* provide a means for distinguishing between the two models.

13.4 The effect of fully concerted subunit interactions

In a fully concerted subunit ineraction model, when the substrate binds to the first subunit, the conformation of the first subunit changes from the F to

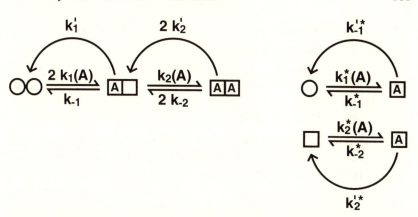

Fig. 13.3. Model of a dimeric enzyme in which the subunits interact in a fully concerted manner. The two intrinsic processes are represented on the right of the figure.

the G conformation, but the subunits interact such that the subunits to which the substrate is not bound also undergo a F to G conformational change. This is illustrated in Figure 13.3. There are two intrinsic processes, one concerned with the binding of the substrate to a subunit in the F conformation and the transition of the subunit from the F to the G conformation. The second intrinsic process is the binding of the substrate to a subunit in the G conformation without transition to another conformation. The relationship between the observed rate constants and the intrinsic rate constants are

$$k_1 = \alpha_{FF} k_1^* \tag{13.44}$$

$$k_{-1} = \alpha_{GG} k_1^* \tag{13.45}$$

$$k_1' = \alpha_{GG} k_1'^* \tag{13.46}$$

$$k_2 = \alpha_{GG} k_2^* \tag{13.47}$$

$$k_{-2} = \alpha_{GG} k_{-2}^* \tag{13.48}$$

$$k_2' = \alpha_{GG} k_2'^* \tag{13.49}$$

The following is the relationships between the observed Michaelis constants and the intrinsic Michaelis constants.

$$K_1 = \frac{\alpha_{GG}}{2\alpha_{FF}} K_1^* \tag{13.50}$$

$$K_2 = 2K_2^* \tag{13.51}$$

The rate equation for the model in Figure 13.3 can be derived by following the procedure employed in the previous sections.

$$v = \frac{2\alpha_{GG}\,E_t(A)[k_1'^*\,K_2^* + k_2'^*(A)]}{K_1^*K_2^* + \dfrac{2\alpha_{FF}K_2^*}{\alpha_{GG}}(A) + \dfrac{\alpha_{FF}}{\alpha_{GG}}(A)^2} \tag{13.52}$$

An analysis of eq. (13.52) shows that the fully concerted subunit interaction model can accommodate all of the shapes that the two previous models can exhibit. However, substrate inhibition is exhibited only if $(1/2) > (k_2'^*/k_1'^*)$.

13.5 The effect of exclusive allosteric subunit interactions

The exclusive allosteric subunit interaction is analogous to the model considered by Monod, Wyman and Changeaux[1], but the symbols employed in this text differ from those employed by Monod, et al. In this model, the free dimer exists in either the FF or GG conformation. The substrate binds to the active site of only the G conformation, and the binding of the substrate does not induce a conformational transition. There are two intrinsic processes, one for the conformational transition of the free dimer and the other for the binding of the substrate and the catalytic reaction. The observed equilibrium constant for the conformational transition is

$$K_1 = \frac{k_{-1}}{k_1} \tag{13.53}$$

This model is presented in Figure 13.4.

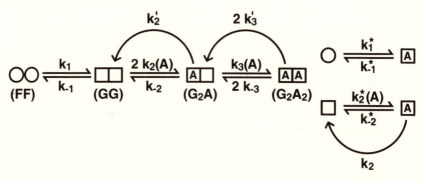

Fig. 13.4. Model of a dimeric enzyme which exhibits exclusive allosteric subunit interaction.

The following are the observed Michaelis constants.

$$K_2^{app} = \frac{k_2 + k_2'}{2k_2} \tag{13.54}$$

$$K_3^{app} = \frac{2(k_{-3} + k_3')}{k_3} \tag{13.55}$$

The following are the relationships between the observed rate constants and the intrinsic rate constants.

$$k_1 = \alpha_{FF} k_1^* \tag{13.56}$$

$$k_{-1} = \alpha_{GG} k_{-1}^* \tag{13.57}$$

$$k_2 = k_3 = \alpha_{GG} k_2^* \tag{13.58}$$

$$k_{-2} = k_{-3} = \alpha_{GG} k_{-2}^* \tag{13.59}$$

$$k_2' = k_3' = \alpha_{GG} k_2'^* \tag{13.60}$$

The total enzyme is

$$E_t = (FF) + (GG) + (G_2A) + (G_2A_2) \tag{13.61}$$

The expressions for the concentration of each enzyme species are

$$(FF) = \frac{\alpha_{GG} K_1^*}{\alpha_{FF}}(GG) \tag{13.62}$$

$$(G_2A) = \frac{2(A)}{K_2^*}(GG) \tag{13.63}$$

$$(G_2A_2) = \frac{(A)^2}{K_2^{*2}}(GG) \tag{13.64}$$

$$(GG) = \frac{K_2^{*2} Et}{K_2^{*2}\left[1 + \frac{\alpha_{GG}}{\alpha_{FF}} K_1^*\right] + 2K_2^*(A) + (A)^2} \tag{13.65}$$

The rate of the enzyme-catalyzed reaction is

$$v = \frac{2\alpha_{GG} k_2'^* E_t(A)(K_2^* + (A)]}{K_2^{*2}\left[1 + \frac{\alpha_{GG}}{\alpha_{FF}} K_1^*\right] + 2K_2^*(A) + (A)^2} \tag{13.66}$$

An analysis of eq. (13.66) shows that the condition necessary for positive cooperativity is $(\alpha_{GG}/\alpha_{FF})K_1^* > 1$. However, the condition that would be necessary if substrate inhibition were to be observed is $1 > 2$. Hence, eq. (13.66) cannot describe substrate inhibition. In like manner, the necessary condition if the substrate-saturation curve is to be hyperbolic is, $1 = 1 + (\alpha_{GG}/\alpha_{FF})K_1^*$ and this is possible only if $K_1^* = 0$. That is, the free enzyme must exist entirely as the GG conformation. Furthermore, the exclusive allosteric subunit interaction model cannot exhibit negative cooperativity. These same conclusions were reported for the effect of subunit interactions on substrate binding[2].

The discussion in this chapter has been restricted to the effect of subunit interaction on the kinetic behavior of a dimeric enzyme. This has been done for conciseness in presentation. The treatment can be extended to higher polymers with little additional difficulty. The study of the steady state kinetic behavior of subunit interaction does not provide the kineticist the opportunity to distinguish between the models studied. The distinction between the models requires investigation of the conformation of the polymer under varying conditions. The kinetic analysis described does provide a basis for expressing a rate equation in terms of kinetic parameters rather than binding constants.

13.6 Problems for chapter 13

13.1   Consider a trimeric enzyme in which the geometry of the subunits is such that each subunit can interact with each other subunit. The model in Figure 13.5 shows the mechanism if each subunit contains an active site and the interaction of subunits is sequential. Derive the rate equation for the foregoing mechanism assuming the concentration of P to be equal to zero.

13.2   What condition must exist if the equation derived in problem 13.1 is to describe a hyperbolic substrate-saturation curve?

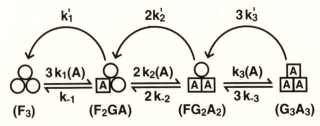

Fig. 13.5. Model of a trimeric enzyme in which the subunits interact in a sequential manner.

13.3 Using the procedure described in this chapter, it can be shown that the condition necessary if a 3:3 rational function is to describe a sigmoidal curve is $\alpha_2\beta_0 > \alpha_1\beta_1$. What conditions must exist if the equation derived in problem 13.1 is to describe a sigmoidal substrate-saturation curve?

13.4 It can also be shown that the condition that must exist if a 3:3 rational function is to describe a curve which shows substrate inhibition is $\alpha_2\beta_3 > \alpha_3\beta_2$. What conditions must exist in the equation derived in problem 13.1 if the equation is to describe a substrate-saturation curve with substrate inhibition?

13.5 Derive the rate equation for the trimer described in problem 13.1 if the subunit interactions were fully concerted.

13.6 What conditions must exist if the equation derived in problem 13.5 were to describe a sigmoidal substrate-saturation curve?

13.7 What conditions would have to exist if the equation derived in problem 13.5 were to describe substrate inhibition?

References

1. Monod, J., Wyman, J. and Changeaux, J. P. (1965). On the nature of allosteric transitions: A plausible model. *J. Mol. Biol.* **12**: 88–118.
2. Koshland, D. E., Nemethy, G. and Filmer, D. (1966). Comparison of experimental binding data and theoretical models in proteins containing subunits. *Biochem.* **5**: 365–85.
3. Ricard, J., Mouttet, C. and Nari, J. (1974). Subunit interactions in enzyme catalysis. *Eur. J. Biochem.* **41**: 479–97.
4. Ricard, J. and Noat, G. (1984). Subunit interactions in enzyme transition states. Antagonism between substrate binding and reaction rates. *J. Theor. Biol.* **111**: 737–53.
5. Ricard, J. and Noat, G. (1985). Subunit coupling and kinetic co-operativity of polymeric enzymes. *J. Theor. Biol.* **117**: 633–49.
6. Ricard, J. and Noat, G. (1986). Catalytic efficiency, kinetic co-operativity of oligomeric enzymes and evolution. *J. Theor. Biol.* **123**: 431–51.

Part Four

Control of multi-enzyme systems

14

Control of linear multi-enzyme systems

The previous sections of this text have discussed the kinetic behavior of individual enzymes in a controlled environment. The need for careful control of the reaction conditions has been stressed. An enormous amount of useful information concerning the mode of action of enzymes has been obtained during this century from these types of studies. Useful as these studies have been, the concept of an enzyme catalyzing a reaction in isolation runs counter to the purpose for which enzymes have been provided in nature. The purpose of an enzyme in nature is to catalyze a reaction in concert with the other enzymes in the metabolic pathway, and the purpose of a metabolic pathway is to catalyze a series of reactions in concert with the many other pathways with which it interacts. Anyone who has given even cursory thought to this matter must have concluded that, in the normal living organism, the action of myriad enzymes is a beautifully coordinated process. On the other hand, if coordination of the action of these enzymes becomes flawed, it is certain that the living organism is going to encounter serious difficulty.

In the view of this author, the investigation of the coordination of multi-enzyme systems is the most exciting challenge to the enzymologist. A multitude of questions present themselves. However, it is essential that the enzymologist follow the example of a judicious detective and ask purposeful questions and interpret the answers obtained in an insightful manner. For example, it is *not* sufficient to ask which enzyme catalyzes a rate-limiting step in a pathway. A more appropriate question is what percent of the control of a pathway does a given enzyme exert under a given set of conditions. The latter question is more appropriate because it recognizes that the control of multi-enzyme systems is a quantitative subject and must be resolved by seeking quantitative answers.

Mathematical treatments of the control of multi-enzyme systems have developed during the past three decades. Higgins appears to have been the first to report a systematic analysis of a sequence of enzymic reactions[1]. Although the symbolism has changed, many of the principles established by Higgins have been incorporated into what has come to be known as metabolic control theory (MCT). The principles of biochemical systems theory (BST) were first reported by Savageau[2-4]. Metabolic control theory was developed by the pioneering work of Kacser and Burns[5] and by Heinrich and Rapoport[6]. Other treatments have been developed[7-10], but the majority of the discussion in this and subsequent chapters will deal with MCT and BST.

14.1 Definition of the parameters of metabolic control

The symbols, terminology and parameters employed in the analysis of multi-enzyme systems differ from those employed in the earlier sections of this text; therefore it is necessary to define the terms which will be utilized in this section. The symbols and terminology will be those employed in MCT. Chapter 16 will contain additional definitions which are employed in BST, but for uniformity, the symbols employed in MCT will be utilized throughout this book. Consider the following linear multi-enzyme pathway. The letter M_i identifies both the intermediate metabolites in the pathway, and the concentration of these metabolites. These are dependent variables. That is, they vary in accordance with the independent variables. The enzymes are denoted by the numerals above the arrows, and they are considered to be parameters of the system. The direction of the arrows indicates the normal direction of flux through the pathway. The direction of the arrow does *not* imply that the reaction is irreversible. The letters X_0 and X_f are external independent variables and are assumed to be outside of the pathway. Thus, X_0 is assumed to be a source of constant size, but this does *not* imply that enzyme 1 is saturated with X_0. It is further assumed that X_f is an infinitely large "sink" so that its fractional concentration does not change significantly. Thus, both the system parameters and the external independent variables are parameters of the conditions described here . In the discussion to follow in this and the succeeding chapters, it will be assumed that only enzymes are involved in the pathway, but in actuality, the pathway may

$$X_0 \xrightarrow{\ \ 1\ \ } M_1 \xrightarrow{\ \ 2\ \ } M_2 \xrightarrow{\ \ 3\ \ } M_3 \xrightarrow{\ \ 4\ \ } X_f$$

Fig. 14.1. Model of a simple linear multi-enzyme system. X_0 is a source of constant size and X_f is an infinitely large sink. M_i are intermediate metabolites.

consist of both enzymes and carriers which transport metabolites across membranes. Carriers exhibit kinetic properties which are analogous to those of enzymes.

The intermediate metabolites are assumed to be in steady state, and they are presumed to be distributed uniformly in a single compartment. Some of these restrictions have been removed in later treatments of MCT. The flux through the pathway is denoted by the letter J. Since the intermediate metabolites are assumed to be in steady state, it follows that $J = v_1 = v_2 = v_3 = v_4$. A final assumption is that a fractional change in the activity of enzyme i results in an identical fractional change in the velocity of the reaction catalyzed by enzyme i. That is, $\partial E_i/E_i = \partial v_i/v_i$.

The following terms are defined in MCT. A fractional change in the flux brought about by a fractional change in the activity of an enzyme is termed the *flux control coefficient*.

$$C_{E_i}^J = \frac{\partial J}{J} \bigg/ \frac{\partial E_i}{E_i} = \frac{E_i}{J}\frac{\partial J}{\partial E_i} = \partial \ln J/\partial \ln E_i \qquad (14.1)$$

Equation (14.1) gives three definitions of the flux control coefficient with respect to enzyme i which are equivalent mathematically. Since there is only one flux involved in the linear pathway portrayed in Fig. 14.1, the flux control coefficient will abbreviated as C_i. The fractional change in the concentration of an intermediate metabolite brought about by a fractional change in the activity of an enzyme is termed as the *concentration control coefficient*.

$$C_{E_i}^{M_j} = \frac{\partial M_j}{M_j} \bigg/ \frac{\partial E_i}{E_i} = \frac{E_i}{M_j}\frac{\partial M_j}{\partial E_i} = \partial \ln M_j/\partial \ln E_i \qquad (14.2)$$

Once again, there are three expressions for the control coefficient which are identical mathematically. Throughout this section, a concentration control coefficient will be abbreviated as $C_i^{M_j}$. The foregoing two control coefficients are properties of the pathway, that is, they are global properties rather than a property of an individual enzyme. It is also important to recognize that the changes must be very small and all the other variables must be kept constant.

There is one more coefficient to be defined, and in the terminology of MCT it is called an *elasticity coefficient*, or more simply, *elasticity*. It is the fractional change in the velocity of an enzymic reaction brought about by a fractional change in a reactant, regardless of whether that reactant be a substrate, or a product, or a modifier produced within the pathway.

$$\varepsilon_{M_j}^{v_j} = \frac{\partial v_j}{v_j} \bigg/ \frac{\partial M_i}{M_i} = \frac{M_i}{v_j} \frac{\partial v_j}{\partial M_i} = \partial \ln v_j / \partial \ln M_i \tag{14.3}$$

The elasticity will be abbreviated ε_i^j. The elasticity of the enzyme toward a reactant is distinct from the control coefficients because it is a local property of the enzyme rather than a global property of the system. Hence, it must be estimated in an investigation of the isolated enzyme, and it is the link between the data obtained by the enzyme kineticist and the data needed for quantitative analysis of the control of a multi-enzyme system. It is useful to consider the elasticity of an enzyme toward its substrate and toward its product in more detail. If M_i is the substrate of enzyme j and M_j is the product of enzyme j, the velocity of the reaction catalyzed by enzyme j is

$$v_j = \frac{V_f[1 - \Gamma/K_{eq}]}{1 + \dfrac{K_{M_i}}{M_i}\left(1 + \dfrac{M_j}{K_{iM_{j_1}}}\right) + \dfrac{M_j}{K_{iM_{j_2}}}} \tag{14.4}$$

where Γ is the mass action ratio, M_j/M_i, and, $K_{iM_{j_1}}$, $K_{iM_{j_2}}$ are product inhibition constants for M_j, and K_{M_i} is the Michaelis constant for M_i. The elasticity of the enzyme toward the substrate is

$$\varepsilon_i^j = \frac{M_i}{v_j} \frac{\partial v_j}{\partial M_i} = \frac{\Gamma/K_{eq}}{1 - \Gamma/K_{eq}} + \frac{\dfrac{K_{M_i}}{M_i}\left[1 + \dfrac{M_j}{K_{iM_{j_1}}}\right]}{1 + \dfrac{K_{M_i}}{M_i}\left[1 + \dfrac{M_j}{K_{iM_{j_1}}}\right] + \dfrac{M_j}{K_{iM_{j_2}}}} \tag{14.5}$$

The right-hand side of eq. (14.5) contains two terms. The first of these is exclusively thermodynamic. If the reaction were at equilibrium the value of this term would be infinity. On the other hand, if the reaction were infinitely far from equilibrium, the value of this thermodynamic term would be 0. The second term on the right-hand side of eq. (14.5) is kinetic. More specifically, it is a measure of the extent to which the enzyme is saturated by the substrate. If the enzyme were saturated by the substrate, the value of this kinetic term would be 0, but, if the enzyme were very far from saturation and the rate equation were a 1:1 function as in eq. (14.4), the value of the kinetic term would be 1. In the case where the rate equation is a higher power rational polynomial, the differentiation necessary to expand the elasticity coefficient would result in a rather complex thermodynamic term, but the maximum value of the kinetic term would reflect the order of the numerator of the rational polynomial. Thus, if the reaction were at equilibrium, the elasticity would be equal to infinity regardless of whether or not the enzyme were saturated with its substrate. However, if the reaction were infinitely far from saturation, the sensitivity of the enzyme to its substrate would depend entirely on the degree to which the enzyme was saturated by the substrate.

The elasticity coefficient for the product can be expanded as follows:

$$\varepsilon_j^j = \frac{M_j}{v_j}\frac{\partial v_j}{\partial M_j} = -\frac{\Gamma/K_{eq}}{1-\Gamma/K_{eq}} - \frac{\dfrac{M_j}{K_{iM_{j_1}}}\left[K_{iM_{j_1}}+\dfrac{K_{M_i}}{M_i}\right]}{1+\dfrac{M_j}{K_{iM_{j_1}}}\left[K_{iM_{j_2}}+\dfrac{K_{M_i}}{M_i}\right]+\dfrac{K_{M_i}}{M_i}} \tag{14.6}$$

Since eq. (14.4) defines the product M_j as an inhibitor, both the terms on the right-hand side of eq. (14.6) are negative. The first term is a measure of the extent to which the reaction has achieved equilibrium while the second term is a measure of the sensitivity of the enzyme to its product.

14.2 Application of sensitivity theory to the control of linear multi-enzyme systems

Cascante et al.[11] have applied the principles of the sensitivity theory to the derivation of expressions for flux and concentration control coefficients in terms of elasticities. These derivations were provided earlier in the development of MCT[5,6,12-15], but the Cascante method presents this derivation in a particularly clear manner. The relationship between the Cascante method and the earlier MCT treatment will be discussed later. The velocity of any reaction in the metabolic pathway portrayed in Figure 14.1 is a function of at least the following variables and parameters.

$$v_i = f(M_1, M_2, M_3, E_1, E_2, E_3, E_4)$$

The flux through the pathway is equal to the velocity of each reaction in the pathway if the intermediate metabolites are in steady state, and, thus the flux and velocity of any reaction are interchangeable. In sensitivity theory, the velocity of each step in the pathway is evaluated for its sensitivity to all the variables and parameters of which it is a function. Thus, the sensitivity of flux to enzyme 1 is given by the following expression.

$$\frac{\partial J}{J}\bigg/\frac{\partial E_1}{E_1} = \left(\frac{\partial v_1}{v_1}\bigg/\frac{\partial M_1}{M_1}\right)\left(\frac{\partial M_1}{M_1}\bigg/\frac{\partial E_1}{E_1}\right) + \left(\frac{\partial v_1}{v_1}\bigg/\frac{\partial M_2}{M_2}\right)\left(\frac{\partial M_2}{M_2}\bigg/\frac{\partial E_1}{E_1}\right)$$

$$+ \left(\frac{\partial v_1}{v_1}\bigg/\frac{\partial M_3}{M_3}\right)\left(\frac{\partial M_3}{M_3}\bigg/\frac{\partial E_1}{E_1}\right) + \left(\frac{\partial v_1}{v_1}\bigg/\frac{\partial E_1}{E_1}\right)$$

$$+\left(\frac{\partial v_1}{v_1}\bigg/\frac{\partial E_2}{E_2}\right)\left(\frac{\partial E_2}{E_2}\bigg/\frac{\partial E_1}{E_1}\right)+\left(\frac{\partial v_1}{v_1}\bigg/\frac{\partial E_3}{E_3}\right)\left(\frac{\partial E_3}{E_3}\bigg/\frac{\partial E_1}{E_1}\right)$$

$$+\left(\frac{\partial v_1}{v_1}\bigg/\frac{\partial E_4}{E_4}\right)\left(\frac{\partial E_4}{E_4}\bigg/\frac{\partial E_1}{E_1}\right) \tag{14.7}$$

Reference to Figure 14.1 indicates that neither M_2 nor M_3 has a direct effect on v_1, therefore, $(\partial v_1/v_1)/(\partial M_2/M_2)$ and $(\partial v_1/v_1)/(\partial M_3/M_3) = 0$. Furthermore, an original assumption of MCT was that there were no enzyme–enzyme interactions and therefore $(\partial E_2/E_2)/(\partial E_1/E_1)$, $(\partial E_3/E_3)/(\partial E_1/E_1)$ and $(\partial E_4 E_4/\partial E_1/E_1) = 0$. The stipulation that there are no enzyme–enzyme interactions has been relaxed in later treatments of MCT[16], and it has never been a required stipulation in BST. However, for simplicity in presentation of the general principles, the assumption that there are no enzyme–enzyme interactions will be employed in the treatment presented in this text. It should be noted that, $C_1 = (\partial J/J)/(\partial E_1/E_1)$, $\varepsilon_1^1 = (\partial v_1/v_1)/(\partial M_1/M_1)$ and $C_1^{M_1} = (\partial M_1/M_1)/(\partial E_1/E_1)$. Since M_1 is the product of v_1, ε_1^1 is negative. To minimize negative signs, throughout this text, any inhibitory elasticity will be defined as $\bar{\varepsilon}_i^j = -\varepsilon_i^j$. It was mentioned previously that a basic assumption inherent in MCT is $(\partial v_1/v_1)/(\partial E_1/E_1) = 1$. However, this assumption is not essential to the sensitivity theory, and so the following definition can be employed $\pi_1^1 = (\partial v_1/v_1)/(\partial E_1 E_1)$. Thus eq. (14.7) can be written as

$$C_1 = -\bar{\varepsilon}_1^1 C_1^{M_1} + \pi_1^1 \tag{14.8}$$

The expressions for the effect of each enzyme on the flux are obtained in the same manner.

$$\frac{\partial J}{J}\bigg/\frac{\partial E_2}{E_2} = \left(\frac{\partial v_2}{v_2}\bigg/\frac{\partial M_1}{M_1}\right)\left(\frac{\partial M_1}{M_1}\bigg/\frac{\partial E_2}{E_2}\right)+\left(\frac{\partial v_2}{v_2}\bigg/\frac{\partial M_2}{M_2}\right)\left(\frac{\partial M_2}{M_2}\bigg/\frac{\partial E_2}{E_2}\right)$$

$$+\left(\frac{\partial v_2}{v_2}\bigg/\frac{\partial M_3}{M_3}\right)\left(\frac{\partial M_3}{M_3}\bigg/\frac{\partial E_2}{E_2}\right)+\left(\frac{\partial v_2}{v_2}\bigg/\frac{\partial E_1}{E_1}\right)\left(\frac{\partial E_1}{E_1}\bigg/\frac{\partial E_2}{E_2}\right)$$

$$+\left(\frac{\partial v_2}{v_2}\bigg/\frac{\partial E_2}{E_2}\right)+\left(\frac{\partial v_2}{v_2}\bigg/\frac{\partial E_3}{E_3}\right)\left(\frac{\partial E_3}{E_3}\bigg/\frac{\partial E_2}{E_2}\right)$$

$$+\left(\frac{\partial v_2}{v_2}\bigg/\frac{\partial E_4}{E_4}\right)\left(\frac{\partial E_4}{E_4}\bigg/\frac{\partial E_2}{E_2}\right) \tag{14.9}$$

$$\frac{\partial J}{J}\bigg/\frac{\partial E_3}{E_3} = \left(\frac{\partial v_3}{v_3}\bigg/\frac{\partial M_1}{M_1}\right)\left(\frac{\partial M_1}{M_1}\bigg/\frac{\partial E_3}{E_3}\right)+\left(\frac{\partial v_3}{v_3}\bigg/\frac{\partial M_2}{M_2}\right)\left(\frac{\partial M_2}{M_2}\bigg/\frac{\partial E_3}{E_3}\right)$$

$$+\left(\frac{\partial v_3}{v_3}\bigg/\frac{\partial M_3}{M_3}\right)\left(\frac{\partial M_3}{M_3}\bigg/\frac{\partial E_3}{E_3}\right)+\left(\frac{\partial v_3}{v_3}\bigg/\frac{\partial E_1}{E_1}\right)\left(\frac{\partial E_1}{E_1}\bigg/\frac{\partial E_3}{E_3}\right)$$

$$+\left(\frac{\partial v_3}{v_3}\bigg/\frac{\partial E_2}{E_2}\right)+\left(\frac{\partial E_2}{E_2}\bigg/\frac{\partial E_3}{E_3}\right)\left(\frac{\partial v_3}{v_3}\bigg/\frac{\partial E_3}{E_3}\right)$$

$$+\left(\frac{\partial v_3}{v_3}\bigg/\frac{\partial E_4}{E_4}\right)\left(\frac{\partial E_4}{E_4}\bigg/\frac{\partial E_3}{E_3}\right) \tag{14.10}$$

$$\frac{\partial J}{J}\bigg/\frac{\partial E_4}{E_4}=\left(\frac{\partial v_4}{v_4}\bigg/\frac{\partial M_1}{M_1}\right)\left(\frac{\partial M_1}{M_1}\bigg/\frac{\partial E_4}{E_4}\right)+\left(\frac{\partial v_4}{v_4}\bigg/\frac{\partial M_2}{M_2}\right)\left(\frac{\partial M_2}{M_2}\bigg/\frac{\partial E_4}{E_4}\right)$$

$$+\left(\frac{\partial v_4}{v_4}\bigg/\frac{\partial M_3}{M_3}\right)\left(\frac{\partial M_3}{M_3}\bigg/\frac{\partial E_4}{E_{E_4}}\right)+\left(\frac{\partial v_4}{v_4}\bigg/\frac{\partial E_1}{E_1}\right)\left(\frac{\partial E_1}{E_1}\bigg/\frac{\partial E_4}{E_4}\right)$$

$$+\left(\frac{\partial v_4}{v_4}\bigg/\frac{\partial E_2}{E_2}\right)\left(\frac{\partial E_2}{E_2}\bigg/\frac{\partial E_4}{E_4}\right)+\left(\frac{\partial v_4}{v_4}\bigg/\frac{\partial E_3}{E_3}\right)\left(\frac{\partial E_3}{E_3}\bigg/\frac{\partial E_4}{E_4}\right)$$

$$+\left(\frac{\partial v_4}{v_4}\bigg/\frac{\partial E_4}{E_4}\right) \tag{14.11}$$

Equations (14.9)–(14.11) can be written in the following form.

$$C_2 = \varepsilon_1^2 C_1^{M_2} - \bar{\varepsilon}_2^2 C_2^{M_2} + \pi_2^2 \tag{14.12}$$

$$C_3 = \varepsilon_2^3 C_3^{M_2} - \bar{\varepsilon}_3^3 C_3^{M_3} + \pi_3^3 \tag{14.13}$$

$$C_4 = \varepsilon_3^4 C_4^{M_3} + \pi_4^4 \tag{14.14}$$

Equations (14.8) and (14.12)–(14.14) constitute a system of equations which can be written in matrix form.

$$\begin{vmatrix}1\\1\\1\\1\end{vmatrix}|C_1\ C_2\ C_3\ C_4|=\begin{vmatrix}-\bar{\varepsilon}_1^1 & 0 & 0\\ \varepsilon_1^2 & -\bar{\varepsilon}_2^2 & 0\\ 0 & \varepsilon_2^3 & -\bar{\varepsilon}_3^3\\ 0 & 0 & \varepsilon_3^4\end{vmatrix}\begin{vmatrix}C_1^{M_1} & C_2^{M_1} & C_3^{M_1} & C_4^{M_1}\\ C_1^{M_2} & C_2^{M_2} & C_3^{M_2} & C_4^{M_2}\\ C_1^{M_3} & C_2^{M_3} & C_3^{M_3} & C_4^{M_3}\end{vmatrix}$$

$$+\begin{vmatrix}\pi_1^1 & 0 & 0 & 0\\ 0 & \pi_2^2 & 0 & 0\\ 0 & 0 & \pi_3^3 & 0\\ 0 & 0 & 0 & \pi_4^4\end{vmatrix}$$

These matrices can be rearranged to give

$$
\begin{vmatrix} 1 \\ 1 \\ 1 \\ 1 \end{vmatrix} |C_1 \ C_2 \ C_3 \ C_4| + \begin{vmatrix} -\bar{\varepsilon}_1^1 & 0 & 0 \\ -\varepsilon_1^2 & \bar{\varepsilon}_2^2 & 0 \\ 0 & -\varepsilon_2^3 & \bar{\varepsilon}_3^3 \\ 0 & 0 & -\varepsilon_3^4 \end{vmatrix} \begin{vmatrix} C_1^{M_1} & C_2^{M_1} & C_3^{M_1} & C_4^{M_1} \\ C_1^{M_2} & C_2^{M_2} & C_3^{M_2} & C_4^{M_2} \\ C_1^{M_3} & C_2^{M_3} & C_3^{M_3} & C_4^{M_3} \end{vmatrix}
$$

$$
= \begin{vmatrix} \pi_1^1 & 0 & 0 & 0 \\ 0 & \pi_2^2 & 0 & 0 \\ 0 & 0 & \pi_3^3 & 0 \\ 0 & 0 & 0 & \pi_4^4 \end{vmatrix}
$$

Matrix algebra provides a means of transforming the rectangular matrices on the left-hand side of the foregoing system of matrices into two square matrices. This is accomplished simply by adding the three columns of the 4×3 matrix onto the column vector of ones, and by adding the vector of flux control coefficients onto the 3×4 matrix of concentration control coefficients. At the same time, the π_i^i elements of the right-hand side of the foregoing matrices can be replaced by 1 because an assumption inherent in MCT is that $(\partial v_i/v_i)/(\partial E_i/E_i) = 1$.

$$
\begin{vmatrix} 1 & \bar{\varepsilon}_1^1 & 0 & 0 \\ 1 & -\bar{\varepsilon}_1^2 & \bar{\varepsilon}_2^2 & 0 \\ 1 & 0 & -\varepsilon_2^3 & \bar{\varepsilon}_3^3 \\ 1 & 0 & 0 & -\varepsilon_3^4 \end{vmatrix} \begin{vmatrix} C_1 & C_2 & C_3 & C_4 \\ C_1^{M_1} & C_2^{M_1} & C_3^{M_1} & C_4^{M_1} \\ C_1^{M_2} & C_2^{M_2} & C_3^{M_2} & C_4^{M_2} \\ C_1^{M_3} & C_2^{M_3} & C_3^{M_3} & C_4^{M_3} \end{vmatrix} = \begin{vmatrix} 1 & 0 & 0 & 0 \\ 0 & 1 & 0 & 0 \\ 0 & 0 & 1 & 0 \\ 0 & 0 & 0 & 1 \end{vmatrix}
$$

These square matrices can be expressed as the following equation.

$$A \times B = I. \tag{14.15}$$

The product of matrices A and B is an identity matrix, and this has important consequences, some of which will be discussed later in this chapter. The significant consequence that will be discussed here is that since an identity matrix is the equivalent of unity in matrix algebra, A is the inverse of B, that is, $A = B^{-1}$, and B is the inverse of A, $B = A^{-1}$. That is what is desired at this point. In chapter 5, the inversion of a matrix was simplified by representing the system as a connection matrix. However, the problem in this case is more complex than the derivation of equations for steady state enzymic reactions. The differential equations for the change of each enzyme species with respect to time are such that all the terms in the enzyme distribution expressions are positive. As a result, it was not necessary to keep track of the sign of the term. In the problem at hand, the terms can be either positive or negative, and thus, it is essential that the sign of each term be evaluated properly. While graph theory has been applied to the derivation of control coefficients[17,18], possibly a more feasible method of derivation is the simplified algorithm for matrix inversion presented in chapter 4 because this algorithm can be incorporated easily into a computer program or it can be followed manually[19].

Each flux and concentration control coefficient will be a quotient, the denominator of which is the symbolic determinant of matrix A in eq. (14.15). The numerator is the symbolic determinant of matrix A in which one of the columns has been replaced by an appropriate column from the identity matrix. Inversion of a matrix is facilitated by the construction of a matrix of non-zero elements. This matrix is

$$Q = \begin{vmatrix} 1 & 2 & 0 \\ 1 & 2 & 3 \\ 1 & 3 & 4 \\ 1 & 4 & 0 \end{vmatrix}$$

To review the procedure briefly, each term in the determinant can be represented by a vector whose elements are selected one from each row of Q in order such that there is no repetition of numbers in the vector. The value of the element is a pointer to the column of matrix A which contains a non-zero element, and the position of the element in the vector indicates the row of A which contains a non-zero element. The sign of the term is $(-1)^p$ where p is the sum of the number of elements out of sequence in the vector plus the number of negative elements selected from matrix A. The vectors, p values as the sum of elements out of sequence plus negative elements, and the actual terms for the denominator determinant for the pathway under consideration are the following:

Vector	p	Term
$(1, 2, 3, 4)$	$0 + 3$	$-\varepsilon_1^2 \varepsilon_2^3 \varepsilon_3^4$
$(2, 1, 3, 4)$	$1 + 2$	$-\bar{\varepsilon}_1^1 \varepsilon_2^3 \varepsilon_3^4$
$(2, 3, 1, 4)$	$2 + 1$	$-\bar{\varepsilon}_1^1 \bar{\varepsilon}_2^2 \varepsilon_3^4$
$(2, 3, 4, 1)$	$3 + 0$	$-\bar{\varepsilon}_1^1 \bar{\varepsilon}_2^2 \bar{\varepsilon}_3^3$

The denominator determinant for all the control coefficients for the pathway portrayed is

$$|D| = -[\varepsilon_1^2 \bar{\varepsilon}_2^3 \varepsilon_3^4 + \bar{\varepsilon}_1^1 \bar{\varepsilon}_2^3 \varepsilon_3^4 + \bar{\varepsilon}_1^1 \bar{\varepsilon}_2^2 \varepsilon_3^4 + \bar{\varepsilon}_1^1 \bar{\varepsilon}_2^2 \bar{\varepsilon}_3^3].$$

The numerator of the flux control coefficient for the first enzyme is obtained by replacing the first column of matrix A of eq. (14.15) with the first column of the identity matrix.

$$\begin{vmatrix} 1 & \bar{\varepsilon}_1^1 & 0 & 0 \\ 0 & -\varepsilon_1^2 & \bar{\varepsilon}_2^2 & 0 \\ 0 & 0 & -\varepsilon_2^3 & \bar{\varepsilon}_3^3 \\ 0 & 0 & 0 & -\varepsilon_3^4 \end{vmatrix} \quad Q = \begin{vmatrix} 1 & 2 \\ 2 & 3 \\ 3 & 4 \\ 4 & 0 \end{vmatrix}$$

The flux control coefficient for enzyme 1 is:

$$C_1 = \varepsilon_1^2 \varepsilon_2^3 \varepsilon_3^4 / |D| = \left[1 + \frac{\bar{\varepsilon}_1^1}{\varepsilon_1^2} + \frac{\bar{\varepsilon}_1^1 \bar{\varepsilon}_2^2}{\varepsilon_1^2 \varepsilon_2^3} + \frac{\bar{\varepsilon}_1^1 \bar{\varepsilon}_2^2 \bar{\varepsilon}_3^3}{\varepsilon_1^2 \varepsilon_2^3 \varepsilon_3^4} \right]^{-1} \tag{14.16}$$

Since the denominator determinant is preceded by a negative sign, it is obvious that the numerator of eq. (14.16) is also negative because the expression for C_1 is positive.

The numerator of the expression for the flux control coefficient for enzyme 2 is obtained by replacing the first column of matrix A with the second column of the identity matrix. The first column of A is replaced by the third column of the identity matrix to obtain the numerator of the flux control coefficient for enzyme 3. In the case of enzyme 4, the first column of A is replaced by the last column of the identity matrix. The following are the flux control coefficients for enzymes 2, 3, and 4, respectively.

$$C_2 = \bar{\varepsilon}_1^1 \varepsilon_2^3 \varepsilon_3^4 / |D| = \left[\frac{\varepsilon_1^2}{\bar{\varepsilon}_1^1} + 1 + \frac{\bar{\varepsilon}_2^2}{\varepsilon_2^3} + \frac{\bar{\varepsilon}_2^2 \bar{\varepsilon}_3^3}{\varepsilon_2^3 \varepsilon_3^4} \right]^{-1} \tag{14.17}$$

$$C_3 = \bar{\varepsilon}_1^1 \bar{\varepsilon}_2^2 \varepsilon_3^4 / |D| = \left[\frac{\varepsilon_1^2 \varepsilon_2^3}{\bar{\varepsilon}_1^1 \bar{\varepsilon}_2^2} + \frac{\varepsilon_2^3}{\bar{\varepsilon}_2^2} + 1 + \frac{\bar{\varepsilon}_3^3}{\varepsilon_3^4} \right]^{-1} \tag{14.18}$$

$$C_4 = \bar{\varepsilon}_1^1 \bar{\varepsilon}_2^2 \bar{\varepsilon}_3^3 / |D| = \left[\frac{\varepsilon_1^2 \varepsilon_2^3 \varepsilon_3^4}{\bar{\varepsilon}_1^1 \bar{\varepsilon}_2^2 \bar{\varepsilon}_3^3} + \frac{\varepsilon_2^3 \varepsilon_3^4}{\bar{\varepsilon}_2^2 \bar{\varepsilon}_3^3} + \frac{\varepsilon_3^4}{\bar{\varepsilon}_3^3} + 1 \right]^{-1} \tag{14.19}$$

The concentration control coefficients for each enzyme with respect to M_1 are obtained by replacing the second column of matrix A with the respective columns of the identity matrix. The concentration control coefficients with respect to M_1 are the following:

$$C_1^{M_1} = (\varepsilon_2^3 \varepsilon_3^4 + \bar{\varepsilon}_2^2 \varepsilon_3^4 + \bar{\varepsilon}_2^2 \bar{\varepsilon}_3^3 / |D| \tag{14.20}$$

$$C_2^{M_1} = -\varepsilon_2^3 \varepsilon_3^4 / |D| \tag{14.21}$$

$$C_3^{M_1} = -\bar{\varepsilon}_2^2 \varepsilon_3^4 / |D| \tag{14.22}$$

$$C_4^{M_1} = -\bar{\varepsilon}_2^2 \bar{\varepsilon}_3^3 / |D| \tag{14.23}$$

In like manner, the concentration control coefficients for each enzyme with respect to M_2 are obtained by replacing the third column of A by the respective columns of the identity matrix.

$$C_1^{M_2} = (\varepsilon_1^2 \varepsilon_3^4 + \varepsilon_1^2 \bar{\varepsilon}_3^3)/|D| \tag{14.24}$$

$$C_2^{M_2} = (\bar{\varepsilon}_1^1 \varepsilon_3^4 + \bar{\varepsilon}_1^1 \bar{\varepsilon}_3^3)/|D| \tag{14.25}$$

$$C_3^{M_2} = -(\varepsilon_1^2 \varepsilon_3^4 + \bar{\varepsilon}_1^1 \varepsilon_3^4)/|D| \tag{14.26}$$

$$C_4^{M_2} = -(\varepsilon_1^2 \bar{\varepsilon}_3^3 + \bar{\varepsilon}_1^1 \bar{\varepsilon}_3^3)/|D| \tag{14.27}$$

Finally, the concentration control coefficient for each enzyme with respect to M_3 can be obtained by replacing the last column of A with each column of the identity matrix.

$$C_1^{M_3} = \varepsilon_1^2 \varepsilon_2^3 / |D| \tag{14.28}$$

$$C_2^{M_3} = \bar{\varepsilon}_1^1 \varepsilon_2^3 / |D| \tag{14.29}$$

$$C_3^{M_3} = \bar{\varepsilon}_1^1 \bar{\varepsilon}_2^2 / |D| \tag{14.30}$$

$$C_4^{M_3} = -(\varepsilon_1^2 \varepsilon_2^3 + \bar{\varepsilon}_1^1 \varepsilon_2^3 + \bar{\varepsilon}_1^1 \bar{\varepsilon}_2^2)/|D| \tag{14.31}$$

14.3 The relationship between sensitivity theory and metabolic control theory

It was stated previously that eq. (14.15) held a number of important consequences. One of these is that since the product of A times B is equal to an identity matrix, the order of multiplication can be reversed. Hence, $A \times B = B \times A = I$. Cascante et al.[11] have pointed out that when the order of multiplication is reversed, the relationships which were basic to the development of MCT are obtained. These basic relationships are called the summation theorem for flux control coefficients, the summation theorems for concentration control coefficients, the connectivity theorems for flux control coefficients and the connectivity theorems for concentration control coefficients. These relationships were essential to derivation of the control coefficients in the original MCT treatments. While these relationships can be obtained by reversing the order of multiplication of A and B in the Cascante treatment, they are not essential to the derivation of expressions for the control coefficients following the principles of sensitivity theory. The multiplication implied is shown in the following:

$$
\begin{vmatrix}
C_1 & C_2 & C_3 & C_4 \\
C_1^{M_1} & C_2^{M_1} & C_3^{M_1} & C_4^{M_1} \\
C_1^{M_2} & C_2^{M_2} & C_3^{M_2} & C_4^{M_2} \\
C_1^{M_3} & C_2^{M_3} & C_3^{M_3} & C_4^{M_3}
\end{vmatrix}
\begin{vmatrix}
1 & \bar{\varepsilon}_1^1 & 0 & 0 \\
1 & -\varepsilon_1^2 & -\bar{\varepsilon}_2^2 & 0 \\
1 & 0 & -\varepsilon_2^3 & \bar{\varepsilon}_3^3 \\
1 & 0 & 0 & -\bar{\varepsilon}_3^4
\end{vmatrix}
=
\begin{vmatrix}
1 & 0 & 0 & 0 \\
0 & 1 & 0 & 0 \\
0 & 0 & 1 & 0 \\
0 & 0 & 0 & 1
\end{vmatrix}
$$

By following the rules of matrix algebra[20], the summation theorems for flux and concentration control coefficients are obtained by pre-multiplication of the rows of B by the first column of A. A brief outline of the rules of matrix multiplication is presented in the appendix of this chapter.

$$C_1 + C_2 + C_3 + C_4 = 1 \tag{14.32}$$

$$C_1^{M_1} + C_2^{M_1} + C_3^{M_1} + C_4^{M_1} = 0 \tag{14.33}$$

$$C_1^{M_2} + C_2^{M_2} + C_3^{M_2} + C_4^{M_2} = 0 \tag{14.34}$$

$$C_1^{M_3} + C_2^{M_3} + C_3^{M_3} + C_4^{M_3} = 0 \tag{14.35}$$

Equation (14.32), the summation theorem for flux control coefficients, states an important principle, namely, that all the enzymes in a pathway contribute to the control of flux through the pathway. It does not mandate that all the enzymes contribute to the same extent. Indeed, in some cases, a flux control coefficient for may be negative, and if such is the case, the sum of the remaining flux control coefficients will exceed unity. That is, there would be amplification of the control exerted by the remaining enzymes. There may be other instances where one enzyme may exert 99 percent of the control of flux through the pathway while the remaining enzymes exert a total of only 1 percent of the control. Whatever the case, the control exerted by any enzyme on the flux through a pathway is a quantitative property of the system and it should be expressed in quantitative terms. The fact that the sum of the control exerted by all the enzymes in the pathway on the concentration of each of the metabolites is equal to zero is a conclusion that could be reached intuitively if the metabolites are in steady state.

The connectivity theorems for the flux control coefficients are obtained by multiplying the first row of B by the second, third and fourth columns of A. The connectivity theorems for the concentration control coefficients are obtained by multiplying each of the last three rows of B by each of the last three columns of A. The resulting equations are required to provide sufficient equations to obtain the relationships expressed in eqs. (14.16)–(14.31) by the original MCT treatments. However, the connectivity theorems are of no particular value to the treatment presented here.

There are some relationships which are of interest and are obtained by carrying out the multiplication implied in eq. (14.15). These are similar to those discussed in the previous paragraph, but they offer an alternative to the repetitive matrix inversion for obtaining the concentration control coefficients. The matrices are

$$
\begin{vmatrix}
1 & \bar{\varepsilon}_1^1 & 0 & 0 \\
1 & -\varepsilon_1^2 & -\bar{\varepsilon}_2^2 & 0 \\
1 & 0 & -\varepsilon_2^3 & \bar{\varepsilon}_3^3 \\
1 & 0 & 0 & -\bar{\varepsilon}_3^4
\end{vmatrix}
\begin{vmatrix}
C_1 & C_2 & C_3 & C_4 \\
C_1^{M_1} & C_2^{M_1} & C_3^{M_1} & C_4^{M_1} \\
C_1^{M_2} & C_2^{M_2} & C_3^{M_2} & C_4^{M_2} \\
C_1^{M_3} & C_2^{M_3} & C_3^{M_3} & C_4^{M_3}
\end{vmatrix}
=
\begin{vmatrix}
1 & 0 & 0 & 0 \\
0 & 1 & 0 & 0 \\
0 & 0 & 1 & 0 \\
0 & 0 & 0 & 1
\end{vmatrix}
$$

Multiplication of the rows of A by the first column of B gives the following relationships.

$$C_1 + \bar{\varepsilon}_1^1 C_1^{M_1} = 1 \tag{14.36}$$

$$C_1 - \varepsilon_2^3 C_1^{M_1} + \bar{\varepsilon}_2^2 C_1^{M_2} = 0 \tag{14.37}$$

$$C_1 - \varepsilon_2^3 C_1^{M_2} + \bar{\varepsilon}_3^3 C_1^{M_3} = 0 \tag{14.38}$$

$$C_1 - \varepsilon_3^4 C_1^{M_3} = 0 \tag{14.39}$$

The value of the foregoing equations lies in the fact that after one has obtained the equations for the flux control coefficient for enzyme 1, the equations for the effect of enzyme 1 on the intermediate metabolites can be obtained by simple algebra without additional matrix inversion. If the expressions are obtained by a computer-based method, this is of little significance. If the equations are derived manually, this can represent a very significant simplification. The right-hand side of eq. (14.36) is 1, but it should be recognized that the 1 can be replaced by the denominator determinant. Relationships relative to enzyme 2 are obtained by multiplying the rows of matrix A by the second column of matrix B.

$$C_2 + \bar{\varepsilon}_1^1 C_2^{M_1} = 0 \tag{14.40}$$

$$C_2 - \varepsilon_1^2 C_2^{M_1} + \bar{\varepsilon}_2^2 C_2^{M_2} = 1 \tag{14.41}$$

$$C_2 - \varepsilon_2^3 C_2^{M_2} + \bar{\varepsilon}_3^3 C_2^{M_3} = 0 \tag{14.42}$$

$$C_2 - \varepsilon_3^4 C_2^{M_3} = 0 \tag{14.43}$$

Multiplying matrix A by the third column of B gives,

$$C_3 + \bar{\varepsilon}_1^1 C_3^{M_1} = 0 \tag{14.44}$$

$$C_3 - \varepsilon_1^2 C_3^{M_1} + \bar{\varepsilon}_2^2 C_3^{M_2} = 0 \tag{14.45}$$

$$C_3 - \varepsilon_2^3 C_3^{M_2} + \bar{\varepsilon}_3^3 C_3^{M_3} = 1 \tag{14.46}$$

$$C_3 - \varepsilon_3^4 C_3^{M_3} = 0 \tag{14.47}$$

The final relationships are obtained by multiplying A with the last column of B.

$$C_4 + \bar{\varepsilon}_1^1 C_4^{M_1} = 0 \tag{14.48}$$

$$C_4 - \varepsilon_1^2 C_4^{M_1} + \bar{\varepsilon}_2^2 C_4^{M_2} = 0 \tag{14.49}$$

$$C_4 - \varepsilon_2^3 C_4^{M_2} + \bar{\varepsilon}_3^3 C_4^{M_3} = 0 \tag{14.50}$$

$$C_4 - \varepsilon_3^4 C_4^{M_3} = 1 \tag{14.51}$$

Equations (14.36)–(14.51) can be summarized as

$$\left[C_i - \sum_{j=1}^{n} \sum_{k=1}^{m} \varepsilon_k^j C_i^{M_k} = \delta_{i,j} \right]_{i=1,2,\dots,n}$$

where n is the number of enzymes, m is the number of intermediate metabolites, and the Kronecker delta, $\delta_{i,j}$, is equal to 1 when $i = j$ and equal to zero when i is not equal to j.

Equations (14.36) through (14.51) provide a simpler method of obtaining the expressions for the concentration control coefficients, but they illustrate some rather profound relationships which may not be apparent intuitively.

It is particularly informative to interpret the eqs. (14.16)–(14.19), the equations for the flux control coefficients, in terms of the expanded expression for the elasticities, namely, eqs. (14.5) and (14.6). It is apparent immediately that if the first enzyme in the pathway is insensitive to inhibition by its product, the first enzyme is the only enzyme which will exert control of the flux through the pathway. A somewhat similar conclusion was expressed years ago in an intuitive manner. That conclusion was often expressed as, "The first committed step in a pathway is the rate controlling step." However, this statement is not entirely accurate. The term "committed step" implies that the step is irreversible, but eq. (14.6) shows that the elasticity of the enzyme for its product may have a non-zero value even though the reaction is infinitely far from equilibrium, for the enzyme may be susceptible to product inhibition. It can be seen that no enzyme downstream from an enzyme which is completely insensitive to inhibition by its product will exert any influence on the flux through the pathway. In like manner, if the reactions in a pathway are all very far from equilibrium, no enzyme upstream from an enzyme which is saturated by its substrate will exert an influence on the flux. Another conclusion which has been expressed often is "An enzymic reaction which is at equilibrium cannot exert any influence on the flux through a pathway". Inspection of eqs. (14.16)–(14.19) provides a mathematical basis for the accuracy of this conclusion. For example, if the second step in the pathway portrayed in Figure 14.1 were at equilibrium, ε_1^2 and $\bar{\varepsilon}_2^2$ would both be equal to infinity. Thus C_2 would be equal to zero.

14.4 The effect of feedback and feed forward loops on the control of a linear pathway

Figure 14.2 portrays a pathway identical to that in Figure 14.1 except that the last intermediate metabolite inhibits the first enzyme in the pathway. Equations (14.12)–(14.14) apply to the pathway in Figure 14.2, but the expression for enzyme 1 is

$$C_1 = -\bar{\varepsilon}_1^1 C_1^{M_1} - \bar{\varepsilon}_3^1 C_1^{M_3} + \pi_1^1 \tag{14.52}$$

The matrices which provide for the derivation of the control coefficients are the following:

$$
\begin{vmatrix}
1 & \bar{\varepsilon}_1^1 & 0 & \bar{\varepsilon}_3^1 \\
1 & -\varepsilon_1^2 & -\bar{\varepsilon}_2^2 & 0 \\
1 & 0 & -\varepsilon_2^3 & \bar{\varepsilon}_3^3 \\
1 & 0 & 0 & -\varepsilon_3^4
\end{vmatrix}
\begin{vmatrix}
C_1 & C_2 & C_3 & C_4 \\
C_1^{M_1} & C_2^{M_1} & C_3^{M_1} & C_4^{M_1} \\
C_1^{M_2} & C_2^{M_2} & C_3^{M_2} & C_4^{M_2} \\
C_1^{M_3} & C_2^{M_3} & C_3^{M_3} & C_4^{M_3}
\end{vmatrix}
=
\begin{vmatrix}
1 & 0 & 0 & 0 \\
0 & 1 & 0 & 0 \\
0 & 0 & 1 & 0 \\
0 & 0 & 0 & 1
\end{vmatrix}
$$

The denominator determinant is

$$|D| = \varepsilon_1^2 \varepsilon_2^3 \varepsilon_3^4 + \bar{\varepsilon}_1^1 \varepsilon_2^3 \varepsilon_3^4 + \bar{\varepsilon}_1^1 \bar{\varepsilon}_2^2 \varepsilon_3^4 + \bar{\varepsilon}_1^1 \bar{\varepsilon}_2^2 \bar{\varepsilon}_3^3 + \varepsilon_1^2 \varepsilon_2^3 \bar{\varepsilon}_3^1 \tag{14.53}$$

The flux control coefficients are

$$C_1 = \varepsilon_1^2 \varepsilon_2^3 \varepsilon_3^4 / |D| = \left[1 + \frac{\bar{\varepsilon}_1^1}{\varepsilon_1^2} + \frac{\bar{\varepsilon}_1^1 \bar{\varepsilon}_2^2}{\varepsilon_1^2 \varepsilon_2^3} + \frac{\bar{\varepsilon}_1^1 \bar{\varepsilon}_2^2 \bar{\varepsilon}_3^3}{\varepsilon_1^2 \varepsilon_2^3 \varepsilon_3^4} + \frac{\bar{\varepsilon}_3^1}{\varepsilon_3^4} \right]^{-1} \tag{14.54}$$

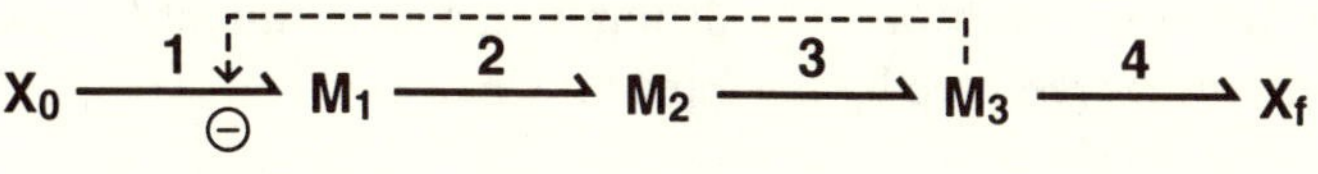

Fig. 14.2. Model of a linear multi-enzyme system with feedback inhibition.

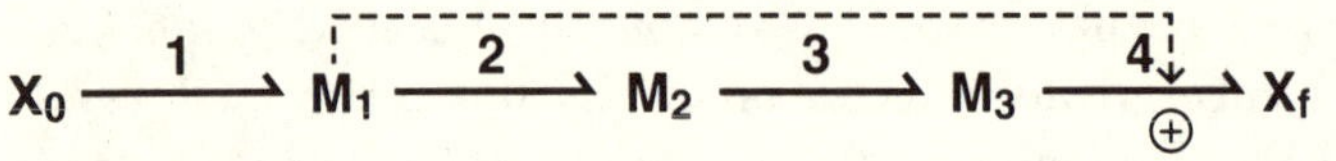

Fig. 14.3. Model of a linear multi-enzyme system with feed forward activation.

$$C_2 = \bar{\varepsilon}_1^1 \varepsilon_2^3 \varepsilon_3^4 / |D| = \left[\frac{\varepsilon_1^2}{\bar{\varepsilon}_1^1} + 1 + \frac{\bar{\varepsilon}_2^2}{\varepsilon_2^3} + \frac{\bar{\varepsilon}_2^2 \bar{\varepsilon}_3^3}{\varepsilon_2^3 \varepsilon_3^4} + \frac{\varepsilon_1^2 \bar{\varepsilon}_3^1}{\bar{\varepsilon}_1^1 \varepsilon_3^4} \right]^{-1} \tag{14.55}$$

$$C_3 = \bar{\varepsilon}_1^1 \bar{\varepsilon}_2^2 \varepsilon_3^4 / |D| = \left[\frac{\varepsilon_1^2 \varepsilon_2^3}{\bar{\varepsilon}_1^1 \bar{\varepsilon}_2^2} + \frac{\varepsilon_2^3}{\bar{\varepsilon}_2^2} + 1 + \frac{\bar{\varepsilon}_3^3}{\varepsilon_3^4} + \frac{\varepsilon_1^2 \varepsilon_2^3 \bar{\varepsilon}_3^1}{\bar{\varepsilon}_1^1 \bar{\varepsilon}_2^2 \varepsilon_3^4} \right]^{-1} \tag{14.56}$$

$$C_4 = (\bar{\varepsilon}_1^1 \bar{\varepsilon}_2^2 \bar{\varepsilon}_3^3 + \varepsilon_1^2 \varepsilon_2^3 \bar{\varepsilon}_3^1) / |D| = \left[\frac{\varepsilon_1^2 \varepsilon_2^3 \varepsilon_3^4 + \bar{\varepsilon}_1^1 \varepsilon_2^3 \varepsilon_3^4 + \bar{\varepsilon}_1^1 \bar{\varepsilon}_2^2 \varepsilon_3^4}{\bar{\varepsilon}_1^1 \bar{\varepsilon}_2^2 \bar{\varepsilon}_3^3 + \varepsilon_1^2 \varepsilon_2^3 \bar{\varepsilon}_3^1} + 1 \right]^{-1} \tag{14.57}$$

It is obvious that feedback inhibition causes a different distribution of control between the enzymes. Of particular interest is the fact that, in contrast to the pathway in Figure 14.1, enzyme 1 does not exert all the control if enzyme 1 is insensitive to product inhibition. If $\bar{\varepsilon}_1^1 = 0$, control of the pathway would be shared between enzymes 1 and 4.

An additional linear pathway will be considered. Figure 14.3 shows a linear pathway in which the first intermediate metabolite functions as an activator of the last enzyme. The expressions for C_1, C_2 and C_3 are as given in eqs. (14.8), (14.12) and (14.13), but the expression for C_4 is given in eq. (14.58).

$$C_4 = \varepsilon_1^4 C_4^{M_1} + \varepsilon_3^4 C_4^{M_3} + \pi_4^4 \tag{14.58}$$

The following are the matrices from which the control coefficients can be derived.

$$\begin{vmatrix} 1 & \bar{\varepsilon}_1^1 & 0 & 0 \\ 1 & -\varepsilon_1^2 & -\bar{\varepsilon}_2^2 & 0 \\ 1 & 0 & -\varepsilon_2^3 & \bar{\varepsilon}_3^3 \\ 1 & -\varepsilon_1^4 & 0 & -\varepsilon_3^4 \end{vmatrix} \begin{vmatrix} C_1 & C_2 & C_3 & C_4 \\ C_1^{M_1} & C_2^{M_1} & C_3^{M_1} & C_4^{M_1} \\ C_1^{M_2} & C_2^{M_2} & C_3^{M_2} & C_4^{M_2} \\ C_1^{M_3} & C_2^{M_3} & C_3^{M_3} & C_4^{M_3} \end{vmatrix} = \begin{vmatrix} 1 & 0 & 0 & 0 \\ 0 & 1 & 0 & 0 \\ 0 & 0 & 1 & 0 \\ 0 & 0 & 0 & 1 \end{vmatrix}$$

With a bit of practice, the foregoing matrices can be constructed directly from the diagram of the pathway. The denominator determinant for the system of equations represented by the foregoing matrices is

$$|D| = \varepsilon_1^2 \varepsilon_2^3 \varepsilon_3^4 + \varepsilon_1^4 \bar{\varepsilon}_2^2 \bar{\varepsilon}_3^3 + \bar{\varepsilon}_1^1 \varepsilon_2^3 \varepsilon_3^4 + \bar{\varepsilon}_1^1 \bar{\varepsilon}_2^2 \varepsilon_3^4 + \bar{\varepsilon}_1^1 \bar{\varepsilon}_2^2 \bar{\varepsilon}_3^3 \tag{14.59}$$

The equations for the flux control coefficients are

$$C_1 = (\varepsilon_1^2 \varepsilon_2^3 \varepsilon_3^4 + \varepsilon_1^4 \bar{\varepsilon}_2^2 \bar{\varepsilon}_3^3)/|D| = \left[1 + \frac{\bar{\varepsilon}_1^1 \varepsilon_2^3 \varepsilon_3^4 + \bar{\varepsilon}_1^1 \bar{\varepsilon}_2^2 \varepsilon_3^4 + \bar{\varepsilon}_1^1 \bar{\varepsilon}_2^2 \bar{\varepsilon}_3^3}{\varepsilon_1^2 \varepsilon_2^3 \varepsilon_3^4 + \varepsilon_1^4 \bar{\varepsilon}_2^2 \bar{\varepsilon}_3^3} \right]^{-1} \tag{14.60}$$

$$C_2 = \bar{\varepsilon}_1^1 \varepsilon_2^3 \varepsilon_3^4/|D| = \left[\frac{\varepsilon_1^2}{\bar{\varepsilon}_1^1} + \frac{\varepsilon_1^4 \bar{\varepsilon}_2^2 \bar{\varepsilon}_3^3}{\bar{\varepsilon}_1^1 \varepsilon_2^3 \varepsilon_3^4} + 1 + \frac{\varepsilon_2^2}{\varepsilon_2^3} + \frac{\bar{\varepsilon}_2^2 \bar{\varepsilon}_3^3}{\varepsilon_2^3 \varepsilon_3^4} \right]^{-1} \tag{14.61}$$

$$C_3 = \bar{\varepsilon}_1^1 \bar{\varepsilon}_2^2 \varepsilon_3^4/|D| = \left[\frac{\varepsilon_1^2 \varepsilon_2^3}{\bar{\varepsilon}_1^1 \bar{\varepsilon}_2^2} + \frac{\varepsilon_1^4 \bar{\varepsilon}_3^3}{\bar{\varepsilon}_1^1 \varepsilon_3^4} + \frac{\varepsilon_2^3}{\varepsilon_2^2} + 1 + \frac{\bar{\varepsilon}_3^3}{\varepsilon_3^4} \right]^{-1} \tag{14.62}$$

$$C_4 = \bar{\varepsilon}_1^1 \bar{\varepsilon}_2^2 \varepsilon_3^3/|D| = \left[\frac{\varepsilon_1^2 \varepsilon_2^3 \varepsilon_3^4}{\bar{\varepsilon}_1^1 \bar{\varepsilon}_2^2 \bar{\varepsilon}_3^3} + \frac{\varepsilon_1^4}{\bar{\varepsilon}_1^1} + \frac{\varepsilon_2^3 \varepsilon_3^4}{\bar{\varepsilon}_2^2 \bar{\varepsilon}_3^3} + \frac{\varepsilon_3^4}{\bar{\varepsilon}_3^3} + 1 \right]^{-1} \tag{14.63}$$

As in the case of the feedback, the feed forward causes a change in the distribution of the control of flux. It is of particular interest that if the third enzyme is insensitive to inhibition by the product, the effect of feed forward is largely abolished. Once again, it is seen that product inhibition can play an important role in the regulation of a metabolic pathway. This possibility was commented upon before the mathematical concepts of metabolic control were developed[20,21]. However, it must be recognized that the elasticity of an enzyme to its product is not due to product inhibition *per se* only, but it also reflects the effect of equilibrium even though the reaction conditions may be rather far from equilibrium.

14.5 The quantitative estimation of control coefficients

The elasticities are local properties of the enzymes, and must be estimated in isolation from the remainder of the pathway. Thus, estimates of elasticities may be obtained from investigations of the kinetic properties of the enzyme provided these studies were conducted under conditions which were comparable to those experienced by the metabolic pathway *in situ*. The control coefficients are global properties of the metabolic pathway, and they must be estimated in the presence of the functioning pathway. Two general techniques have been employed in these studies, and they involve either titration of an enzyme with an inhibitor which binds tightly to the enzyme or by genetic manipulation of enzyme activity. In these studies, flux or metabolite concentration is plotted against enzyme activity[22-26]. A tangent is drawn at some point in the curve, and the control coefficient at that point is equal to the slope of the tangent times a weighting factor which is E_i/J in the case of a flux control coefficient, or E_i/M_j in the case of a concentration control coefficient. A much more direct measurement of a control coeffi-

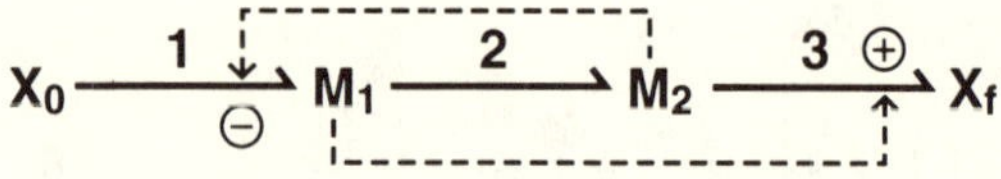

Fig. 14.4. A linear multi-enzyme system with both feedback inhibition and feed forward activation.

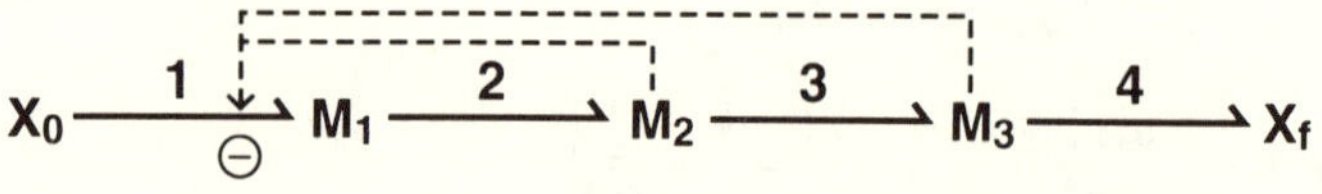

Fig. 14.5. A linear multi-enzyme system with two-fold feedback inhibition.

cient is a plot of flux or metabolite concentration against enzyme activity in log-log space. In the latter case, the control coefficient is simply the slope of the plot at any point. Considerable progress has been made in an attempt to develop feasible methods for estimating control coefficients[27-32].

14.6 Problems for chapter 14

14.1   Figure 14.4 presents a multi-enzyme pathway. Use the procedures described in this chapter to obtain the matrices necessary to derive the control coefficients. Assume that $\pi_i^i = 1$.

14.2   Multiply the A matrix obtained in problem 14.1 by the B matrix to obtain the relationships between the flux control coefficients and the concentration control coefficients, and then reverse the order of multiplication of these matrices to obtain the summation and connectivity relationships of MCT.

14.3   Derive expressions for the flux and concentration control coefficients for the pathway portrayed in problem 14.1.

14.4   Figure 14.5 presents a multi-enzyme pathway. Derive the expressions for the flux and concentration control coefficients for the foregoing pathway.

Appendix

An important operation in matrix algebra is the multiplication of matrices. Consider the following system of matrices.

$$
\begin{vmatrix} a_{11} & a_{12} & a_{13} \\ a_{21} & a_{22} & a_{23} \\ a_{31} & a_{32} & a_{33} \end{vmatrix}
\begin{vmatrix} b_{11} & b_{12} & b_{13} \\ b_{21} & b_{22} & b_{23} \\ b_{31} & b_{32} & b_{33} \end{vmatrix}
=
\begin{vmatrix} c_{11} & c_{12} & c_{13} \\ c_{21} & c_{22} & c_{23} \\ c_{31} & c_{32} & c_{33} \end{vmatrix}
$$

The foregoing system can be abbreviated

$A \times B = C$

The elements of C are each equal to the sum of the multiplication of each row of A by a column of B according to the following procedure.

$c_{11} = a_{11}b_{11} + a_{12}b_{21} + a_{13}b_{31}$
$c_{21} = a_{21}b_{11} + a_{22}b_{21} + a_{23}b_{31}$
$c_{31} = a_{31}b_{11} + a_{32}b_{21} + a_{33}b_{31}$

$c_{12} = a_{11}b_{12} + a_{12}b_{22} + a_{13}b_{32}$
$c_{22} = a_{21}b_{12} + a_{22}b_{22} + a_{23}b_{32}$
$c_{23} = a_{31}b_{12} + a_{32}b_{22} + a_{33}b_{32}$

$c_{13} = a_{11}b_{13} + a_{12}b_{23} + a_{13}b_{33}$
$c_{23} = a_{21}b_{13} + a_{22}b_{23} + a_{23}b_{33}$
$c_{33} = a_{31}b_{13} + a_{32}b_{23} + a_{33}b_{33}$

In general, a different value is obtained for the elements of C if the order of multiplication of A and B is reversed. However, if C is an identity matrix, as is the case in many of the instances covered in this book, the order of multiplication of A and B can be reversed. The reason for this is that an identity matrix is the equivalent of unity in matrix algebra, and in that case $A = B^{-1}$ and $B = A^{-1}$.

References

1. Higgins, J. (1963). Analysis of sequential reactions. *Ann. N. Y. Acad. Sci.* **108**: 305–21.
2. Savageau, M. A. (1969). Biochemical systems analysis. I. Some mathematical properties of the rate law for the component enzymatic reactions. *J. Theor. Biol.* **25**: 365–69.
3. Savageau, M. A. (1969). Biochemical systems analysis. II. The steady-state solutions for an *n*-pool system using power law approximation. *J. Theor. Biol.* **25**: 370–79.
4. Savageau, M. A. (1970). Biochemical systems analysis. III Dynamic solutions using a power-law approximation. *J. Theor. Biol.* **26**: 215–26.
5. Kacser, H. and Burns, J. A. (1972). The control of flux. *Symp. Soc. Exp. Biol.* **27**: 65–104.
6. Heinrich, R. and Rapoport, T. A. (1974). A linear steady-state treatment of enzymatic chains. *Eur. J. Biochem.* **42**: 89–95.
7. Crabtree, B. and Newsholme, E. A. (1985). A quantitative approach to metabolic control. *Curr. Top. Cell. Regul.* **25**: 21–76.
8. Welch, G. R., Keleti, T. and Vertessy, B. (1988). The control of cell metabolism for homogeneous and heterogeneous enzyme systems. *J. Theor. Biol.* **130**: 407–22.
9. Kohn, M. C. and Letzkus, W. J. (1983). A graph- theoretical analysis of metabolic regulation. *J. Theor. Biol.* **100**: 293–304.
10. Thomas, K. (1991). Regulatory networks seen as asynchronous automata: A logical description. *J. Theor. Biol.* **153**: 1–23.
11. Cascante, M., Franco, R. and Canela, E. I. (1989). Use of implicit methods from general sensitivity theory to develop a systematic approach to metabolic control. I Unbranched pathways. *Math. Biosci.* **94**: 271–88.

12. Kacser, H. (1983). The control of enzyme systems *in vivo*: Elasticity analysis of the steady state. *Biochem. Soc. Trans.* **11**: 35–40.
13. Westerhoff, H. V. and Chen, Y-D. (1984). How do enzyme activities control metabolite concentrations? *Eur. J. Biochem.* **142**: 425–30.
14. Fell, D. A. and Sauro, H. M. (1985). Metabolic control and its analysis. Additional relationships between elasticities and control coefficients. *Eur. J. Biochem.* **148**: 555–61.
15. Sauro, H. M., Small, J. R. and Fell, D. A. (1987). Metabolic control and its analysis. Extensions to the theory and matrix method. *Eur. J. Biochem.* **165**: 215–221.
16. Kacser, H., Sauro, H. M. and Acerenza, L. (1990). Enzyme-enzyme interactions and control analysis. *Eur. J. Biochem.* **187**: 481–91.
17. Sen, A. K. (1990). Metabolic control analysis. An application of signal flow graphs. *Biochem. J.* **269**: 141–47.
18. Sen, A. K. (1990). Topological analysis of metabolic control. *Math. Biosci.* **102**: 191–223.
19. Schulz, A. R. (1991). Algorithms for the derivation of flux and concentration control coefficients. *Biochem. J.* **278**: 299–304.
20. Walter, C. and Frieden, E. (1963). The prevalence and significance of product inhibition of enzymes. *Adv. Enzymol.* **25**: 167–274.
21. Koch, A. L. (1967). Metabolic control through reflexive enzyme action. *J. Theor. Biol.* **15**: 75–102.
22. Westerhoff, H. V. and Arents, J. C. (1984). Two (completely) rate-limiting steps in one metabolic pathway *Biosci. Reports* **4**: 23–31.
23. Salter, M., Knowles, R. G. and Pogson, C. I. (1986). Quantification of the importance of individual steps in the control of aromatic amino acid metabolism. *Biochem. J.* **234**: 635–47.
24. Fell, D. A. and Snell, K. (1987). Control analysis of mammalian serine biosynthesis. *Biochem. J.* **256**: 97–101.
25. Westerhoff, H. V., Plomp, P. J. A. M., Groen, A. K., Wanders, R. J. A., Bode, J. A. and Van Dam, K. (1987). On the origin of the limited control of mitochondrial respiration by the adenine nucleotide translocator. *Arch. Biochem. Biophys.* **257**: 154–69.
26. Page, R. A., Kitson, K. E. and Hardman, H. J. (1991). The importance of alcohol dehydrogenase in regulation of ethanol metabolism in rat liver cells. *Biochem. J.* **278**: 659–65.
27. Brown, G. C., Hafner, R. P. and Brand, M. D. (1990). A 'top-down' approach to the determination of control coefficients in metabolic control theory. *Eur. J. Biochem.* **188**: 321–25.
28. Kahn, D. and Westerhoff, H. V. (1991). Control theory of regulatory cascades. *J. Theor. Biol.* **153**: 255–85.
29. Delgado, J. P. and Liao, J. C. (1991). Identifying rate-controlling enzymes in metabolic pathways without kinetic parameters. *Biotechnol. Prog.* **7**: 15–20.
30. Delgado, J. and Liao, J. C. (1992). Determination of flux control coefficients from transient metabolite concentrations. *Biochem. J.* **282**: 919–27.
31. Delgado, J. and Liao, J. C. (1992). Metabolic control analysis using transient metabolite concentrations. *Biochem. J.* **285**: 965–72.
32. Quant, P. A. (1993). Experimental application of 'top-down' control analysis to metabolic sytems. *Trends Biochem. Sci.* **18**: 26–30.

15

Control of branched multi-enzyme systems

The control of flux through linear pathways was discussed in the previous chapter, but many, if not most, metabolic pathways contain points at which other branches either converge or diverge from the pathway. La Porte et al.[1] wrote, "Perhaps the most pervasive type of cellular control is the metering of flux between competing pathways." This point was stressed earlier by Holzer[2]. In this chapter, it will be shown that flux through one branch can amplify the control exerted by enzymes in another branch. This phenomenon can be particularly striking in the case of substrate cycles where one branch operates in direct opposition to the other branch of the pathway, and it presents a function other than a futile cycle for substrate cycles.

15.1 Application of the sensitivity theory to branched pathways

Consider the following branched multi-enzyme pathway. The branch diverges from metabolite 2 in the main pathway portrayed in Figure 15.1. Enzyme 2.1 catalyzes the divergent branch. There are three fluxes involved in the pathway. Flux J_1 is the flux prior to the branch point, flux J_2 is the flux through the divergent branch, and J_3 is the flux through the main path after the branch point. Cascante et al.[3] extended the general sensitivity theory to branched metabolic pathways, and their treatment will be presented here.

The following relationship between the three fluxes is obvious from Figure 15.1.

$$J_1 = J_2 + J_3 \tag{15.1}$$

Differentiation of eq. (15.1) with respect to any enzyme in the pathway when all other variables are held constant gives

$$\frac{\partial J_1}{\partial E_i} = \frac{\partial J_2}{\partial E_i} + \frac{\partial J_3}{\partial E_i} \tag{15.2}$$

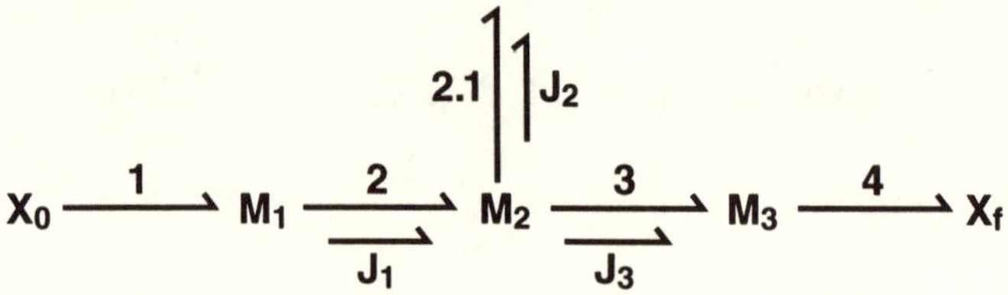

Fig. 15.1. Model of a multi-enzyme system with a single divergent branch.

Equation (15.2) can be normalized and expressed as

$$J_1 C_i^1 = J_2 C_i^2 + J_3 C_i^3 \tag{15.3}$$

where $C_i^j = C_{E_i}^{J_j}$. Thus,

$$C_i^1 = \hat{J}_2 C_i^2 + \hat{J}_3 C_i^3 \tag{15.4}$$

where $\hat{J}_i = J_i / J_1$. Hence, all the fluxes are considered relative to flux 1. Actually, anyone of the three fluxes could be employed as the reference flux. The equations arising from the sensitivity theory for the pathway in Figure 15.1 are

$$C_1^1 = \hat{J}_2 C_1^2 + \hat{J}_3 C_1^3 = -\bar{\varepsilon}_1^1 C_1^{M_1} + \pi_1^1 \tag{15.5}$$

$$C_2^1 = \hat{J}_2 C_2^2 + \hat{J}_3 C_2^3 = -\varepsilon_1^2 C_2^{M_1} - \bar{\varepsilon}_2^2 C_2^{M_2} + \pi_2^2 \tag{15.6}$$

$$C_{2.1}^2 = \varepsilon_2^{2.1} C_{2.1}^{M_2} + \pi_{2.1}^{2.1} \tag{15.7}$$

$$C_3^3 = \varepsilon_2^3 C_3^{M_2} - \bar{\varepsilon}_3^3 C_3^{M_3} + \pi_3^3 \tag{15.8}$$

$$C_4^3 = \varepsilon_3^4 C_4^{M_3} + \pi_4^4 \tag{15.9}$$

Equations (15.5)–(15.9) can be expressed in matrix form.

$$
\begin{vmatrix}
\hat{J}_2 & \hat{J}_3 \\
\hat{J}_2 & \hat{J}_3 \\
1 & 0 \\
0 & 1 \\
0 & 1
\end{vmatrix}
\begin{vmatrix}
C_1^2 & C_2^2 & C_{2.1}^2 & C_3^2 & C_4^2 \\
C_1^3 & C_2^3 & C_{2.1}^3 & C_3^3 & C_4^3
\end{vmatrix}
$$

$$
=
\begin{vmatrix}
-\bar{\varepsilon}_1^1 & 0 & 0 \\
\varepsilon_1^2 & \bar{\varepsilon}_2^2 & 0 \\
0 & \varepsilon_2^{2.1} & 0 \\
0 & \varepsilon_2^3 & -\bar{\varepsilon}_3^3 \\
0 & 0 & \varepsilon_3^4
\end{vmatrix}
\begin{vmatrix}
C_1^{M_1} & C_2^{M_1} & C_{2.1}^{M_1} & C_3^{M_1} & C_4^{M_1} \\
C_1^{M_2} & C_2^{M_2} & C_{2.1}^{M_2} & C_3^{M_2} & C_4^{M_2} \\
C_1^{M_3} & C_2^{M_3} & C_{2.1}^{M_3} & C_3^{M_3} & C_5^{M_3}
\end{vmatrix}
\begin{vmatrix}
\pi_1^1 & 0 & 0 & 0 & 0 \\
0 & \pi_2^2 & 0 & 0 & 0 \\
0 & 0 & \pi_{2.1}^{2.1} & 0 & 0 \\
0 & 0 & 0 & \pi_3^3 & 0 \\
0 & 0 & 0 & 0 & \pi_4^4
\end{vmatrix}
$$

Since the assumption $\pi_i^i = (\partial v_i/v_i)/(\partial E_i/E_i) = 1$ is inherent in the treatment presented in this text, the previous matrices can be rearranged

$$
\begin{vmatrix}
\hat{J}_2 & \hat{J}_3 & \bar{\varepsilon}_1^1 & 0 & 0 \\
\hat{J}_2 & \hat{J}_3 & -\varepsilon_1^2 & \bar{\varepsilon}_2^2 & 0 \\
1 & 0 & 0 & -\varepsilon_2^{2.1} & 0 \\
0 & 1 & 0 & -\varepsilon_2^3 & \bar{\varepsilon}_3^3 \\
0 & 1 & 0 & 0 & -\varepsilon_3^4
\end{vmatrix}
\begin{vmatrix}
C_1^2 & C_2^2 & C_{2.1}^2 & C_3^2 & C_4^2 \\
C_1^3 & C_2^3 & C_{2.1}^3 & C_3^3 & C_4^3 \\
C_1^{M_1} & C_2^{M_1} & C_{2.1}^{M_1} & C_3^{M_1} & C_4^{M_1} \\
C_1^{M_2} & C_2^{M_2} & C_{2.1}^{M_2} & C_3^{M_2} & C_4^{M_2} \\
C_1^{M_3} & C_2^{M_3} & C_{2.1}^{M_3} & C_3^{M_3} & C_4^{M_3}
\end{vmatrix}
=
\begin{vmatrix}
1 & 0 & 0 & 0 & 0 \\
0 & 1 & 0 & 0 & 0 \\
0 & 0 & 1 & 0 & 0 \\
0 & 0 & 0 & 1 & 0 \\
0 & 0 & 0 & 0 & 1
\end{vmatrix}
$$

The above matrices represent the following equation:

$$A \times B = I \tag{15.10}$$

The summation theorems of MCT are obtained by pre-multiplying the first two columns of matrix A by matrix B. The summation theorem for the flux control coefficients are

$$\hat{J}_2 C_1^2 + \hat{J}_2 C_2^2 + C_{2.1}^2 = 1 \tag{15.11}$$

$$\hat{J}_3 C_1^2 + \hat{J}_3 C_2^2 + C_3^2 + C_4^2 = 0 \tag{15.12}$$

$$\hat{J}_2 C_1^3 + \hat{J}_2 C_2^3 + C_{2.1}^3 = 0 \tag{15.13}$$

$$\hat{J}_3 C_1^3 + \hat{J}_3 C_2^3 + C_3^3 + C_4^3 = 1 \tag{15.14}$$

The complete summation theorem for flux 2 is given by the sum of eqs. (15.11) and (15.12), and the sum of these equations is unity. Likewise, the summation theorem for flux 3 is the sum of eqs. (15.13) and (15.14). However, eqs. (15.12) and (15.13) impose additional constraints. It is obvious that either all the terms in the latter two expressions must be equal to zero, or else one or more of the terms in each of the equations must be negative. If the latter is true, there is amplification of the control exerted by the non-negative terms. Thus, branching in the pathway can result in amplification of the control exerted by some of the enzymes.

The denominator determinant for the control coefficients is obtained by an inversion of matrix A of eq. (15.10). Since this is a larger matrix than those dealt with in earlier chapters, the inversion will be presented here using the Q matrix to facilitate the process.

$$
A =
\begin{vmatrix}
\hat{J}_2 & \hat{J}_3 & \bar{\varepsilon}_1^1 & 0 & 0 \\
\hat{J}_2 & \hat{J}_3 & -\varepsilon_1^2 & \bar{\varepsilon}_2^2 & 0 \\
1 & 0 & 0 & -\varepsilon_2^{2.1} & 0 \\
0 & 1 & 0 & -\varepsilon_2^3 & \bar{\varepsilon}_3^3 \\
0 & 1 & 0 & 0 & -\varepsilon_3^4
\end{vmatrix}
\qquad
Q =
\begin{vmatrix}
1 & 2 & 3 & 0 \\
1 & 2 & 3 & 4 \\
1 & 4 & 0 & 0 \\
2 & 4 & 5 & 0 \\
2 & 5 & 0 & 0
\end{vmatrix}
$$

The terms which comprise the denominator determinant are as follows: The first column is the vector representation of the term, the second column is the p value, and the last column actual term.

$(1,3,4,2,5)$ $2+3$ $-\hat{J}_2 \varepsilon_1^2 \varepsilon_2^{2\cdot 1} \varepsilon_3^4$

$(1,3,4,5,2)$ $3+2$ $-\hat{J}_2 \varepsilon_1^2 \varepsilon_2^{2\cdot 1} \bar{\varepsilon}_3^4$

$(2,3,1,4,5)$ $2+3$ $-\hat{J}_3 \varepsilon_2^2 \varepsilon_3^3 \varepsilon_3^4$

$(3,1,4,2,5)$ $3+2$ $-\hat{J}_2 \bar{\varepsilon}_1^1 \varepsilon_2^{2\cdot 1} \varepsilon_3^4$

$(3,1,4,5,2)$ $4+1$ $-\hat{J}_2 \bar{\varepsilon}_1^1 \varepsilon_2^{2\cdot 1} \bar{\varepsilon}_3^3$

$(3,2,1,4,5)$ $3+2$ $-\hat{J}_3 \bar{\varepsilon}_1^1 \varepsilon_2^3 \varepsilon_3^4$

$(3,4,1,2,5)$ $4+1$ $-\bar{\varepsilon}_1^1 \bar{\varepsilon}_2^2 \varepsilon_3^4$

$(3,4,1,5,2)$ $5+0$ $-\bar{\varepsilon}_1^1 \bar{\varepsilon}_2^2 \bar{\varepsilon}_3^3$

Thus, the symbolic denominator determinant is

$$|D| = -[\hat{J}_2(\varepsilon_1^2 + \bar{\varepsilon}_1^1)\varepsilon_2^{2\cdot 1}(\varepsilon_3^4 + \bar{\varepsilon}_3^3) + \hat{J}_3(\varepsilon_1^2 + \bar{\varepsilon}_1^1)\varepsilon_2^3 \varepsilon_3^4 + \bar{\varepsilon}_1^1 \bar{\varepsilon}_2^2(\varepsilon_3^4 + \bar{\varepsilon}_3^3)] \quad (15.15)$$

The flux control coefficients for flux 2 are given by the following

$$C_1^2 = \varepsilon_1^2 \varepsilon_2^{2\cdot 1}(\varepsilon_3^4 + \bar{\varepsilon}_3^3)/|D|$$

$$= \left[\frac{\hat{J}_2(\varepsilon_1^2 + \bar{\varepsilon}_1^1)}{\varepsilon_1^2} + \frac{\hat{J}_3(\varepsilon_1^2 + \bar{\varepsilon}_1^1)\varepsilon_2^3 \varepsilon_3^4}{\varepsilon_1^2 \varepsilon_2^{2\cdot 1}(\varepsilon_3^4 + \bar{\varepsilon}_3^3)} + \frac{\bar{\varepsilon}_1^1 \bar{\varepsilon}_2^2}{\varepsilon_1^2 \varepsilon_2^{2\cdot 1}} \right]^{-1} \quad (15.16)$$

$$C_2^2 = \bar{\varepsilon}_1^1 \varepsilon_2^{2\cdot 1}(\varepsilon_3^4 + \bar{\varepsilon}_3^3)/|D|$$

$$= \left[\frac{\hat{J}_2(\varepsilon_1^2 + \bar{\varepsilon}_1^1)}{\bar{\varepsilon}_1^1} + \frac{\hat{J}_3(\varepsilon_1^2 + \bar{\varepsilon}_1^1)\varepsilon_2^3 \varepsilon_3^4}{\bar{\varepsilon}_1^1 \varepsilon_2^{2\cdot 1}(\varepsilon_3^4 + \bar{\varepsilon}_3^3)} + \frac{\bar{\varepsilon}_2^2}{\varepsilon_2^{2\cdot 1}} \right]^{-1} \quad (15.17)$$

$$C_{2.1}^2 = [\hat{J}_3(\varepsilon_1^2 + \bar{\varepsilon}_1^1)\varepsilon_2^3 \varepsilon_3^4 + \bar{\varepsilon}_1^1 \bar{\varepsilon}_2^2(\varepsilon_3^4 + \bar{\varepsilon}_3^3)/|D|$$

$$= \left[\frac{\hat{J}_2(\varepsilon_1^2 + \bar{\varepsilon}_1^1)\varepsilon_2^{2\cdot 1}(\varepsilon_3^4 + \bar{\varepsilon}_3^3) + \hat{J}_3(\varepsilon_1^2 + \bar{\varepsilon}_1^1)\varepsilon_2^3 \varepsilon_3^4 + \bar{\varepsilon}_2^1 \bar{\varepsilon}_2^2(\varepsilon_3^4 + \bar{\varepsilon}_3^3)}{\hat{J}_3(\varepsilon_1^2 + \bar{\varepsilon}_1^1)\varepsilon_2^3(\varepsilon_3^4 + \bar{\varepsilon}_3^3)} \right]^{-1} \quad (15.18)$$

$$C_3^2 = -\hat{J}_3(\varepsilon_1^2 + \bar{\varepsilon}_1^1)\varepsilon_2^{2\cdot 1} \varepsilon_3^4/|D|$$

$$= -\left[\frac{\hat{J}_2(\varepsilon_1^2 + \bar{\varepsilon}_1^1)}{\hat{J}_3 \varepsilon_3^4} + \frac{\varepsilon_2^3}{\varepsilon_2^{2\cdot 1}} + \frac{\bar{\varepsilon}_1^1 \bar{\varepsilon}_2^2(\varepsilon_3^4 + \bar{\varepsilon}_3^3)}{(\varepsilon_1^2 + \bar{\varepsilon}_1^1)\varepsilon_2^{2\cdot 1} \bar{\varepsilon}_3^3} \right]^{-1} \quad (15.19)$$

$$C_4^2 = -\hat{J}_3(\varepsilon_1^2 + \bar{\varepsilon}_1^1)\varepsilon_2^{2\cdot 1} \bar{\varepsilon}_3^3/|D|$$

$$= -\left[\frac{\hat{J}_2(\varepsilon_1^2 + \bar{\varepsilon}_1^1)}{\hat{J}_3 \bar{\varepsilon}_3^3} + \frac{\varepsilon_2^3}{\varepsilon_2^{2\cdot 1}} + \frac{\bar{\varepsilon}_1^1 \bar{\varepsilon}_2^2(\varepsilon_3^4 + \bar{\varepsilon}_3^3)}{(\varepsilon_1^2 + \bar{\varepsilon}_1^1)\varepsilon_2^{2\cdot 1} \bar{\varepsilon}_3^3} \right]^{-1} \quad (15.20)$$

As shown in eq. (15.15), the denominator determinant is preceded by a minus sign. Hence, the positive flux control coefficients have negative numerators while the negative flux control coefficients have positive numerators.

A number of features are apparent from the foregoing equations for flux control coefficients. If the enzyme 2.1, the first enzyme in branch 2, were saturated by the branch point metabolite, the enzymes in the main pathway would exert no control on the flux through branch 2. It should be recalled that in a linear pathway without a feedback loop no enzyme 'downstream' from an enzyme which was insensitive to product inhibition exerted any control on the flux through the pathway. This is true of the branched pathway shown in Figure 15.1 prior to the branch point, but it is *not* true after the branch point.

The following are the flux control coefficients for flux 3

$$C_1^3 = \varepsilon_1^2 \varepsilon_2^3 \varepsilon_3^4 / |\mathbf{D}|$$

$$= \left[\frac{\hat{J}_3(\varepsilon_1^2 + \bar{\varepsilon}_1^1)}{\varepsilon_1^2} + \frac{[\hat{J}_2(\varepsilon_1^2 + \bar{\varepsilon}_1^1)\varepsilon_2^{2\cdot 1} + \bar{\varepsilon}_1^1 \bar{\varepsilon}_2^2](\varepsilon_3^4 + \bar{\varepsilon}_3^3)}{\varepsilon_1^2 \varepsilon_2^3 \varepsilon_3^4} \right]^{-1} \tag{15.21}$$

$$C_2^3 = \bar{\varepsilon}_1^1 \varepsilon_2^3 \varepsilon_3^4 / |\mathbf{D}|$$

$$= \left[\frac{\hat{J}_3(\varepsilon_1^2 + \bar{\varepsilon}_1^1)}{\bar{\varepsilon}_1^1} + \frac{[\hat{J}_2(\varepsilon_1^2 + \bar{\varepsilon}_1^1)\varepsilon_2^{2\cdot 1} + \bar{\varepsilon}_1^1 \bar{\varepsilon}_2^2](\varepsilon_3^4 + \bar{\varepsilon}_3^3)}{\bar{\varepsilon}_1^1 \varepsilon_2^3 \varepsilon_3^4} \right]^{-1} \tag{15.22}$$

$$C_{2.1}^3 = -\hat{J}_2(\varepsilon_1^2 + \bar{\varepsilon}_1^1)\varepsilon_2^3 \varepsilon_3^4 / |\mathbf{D}|$$

$$= -\left[\frac{\hat{J}_3}{\hat{J}_2} \frac{\varepsilon_2^{2\cdot 1}(\varepsilon_3^4 + \bar{\varepsilon}_3^3)}{\varepsilon_2^3 \varepsilon_3^4} + \frac{\bar{\varepsilon}_1^1 \bar{\varepsilon}_2^2(\varepsilon_3^4 + \bar{\varepsilon}_3^3)}{\hat{J}_2(\varepsilon_1^2 \bar{\varepsilon}_1^1)\varepsilon_2^3 \varepsilon_3^4} \right]^{-1} \tag{15.23}$$

$$C_3^3 = [\hat{J}_2(\varepsilon_1^2 + \bar{\varepsilon}_1^1)\varepsilon_2^{2\cdot 1} + \bar{\varepsilon}_1^1 \bar{\varepsilon}_2^2]\varepsilon_3^4 / |\mathbf{D}|$$

$$= \left[\frac{\hat{J}_3(\varepsilon_1^2 + \bar{\varepsilon}_1^1)\varepsilon_2^3}{\hat{J}_2(\varepsilon_1^2 + \bar{\varepsilon}_1^1)\varepsilon_2^{2\cdot 1} + \bar{\varepsilon}_1^1 \bar{\varepsilon}_2^2} + \frac{\varepsilon_3^4 + \bar{\varepsilon}_3^3}{\varepsilon_3^4} \right]^{-1} \tag{15.24}$$

$$C_4^3 = [\hat{J}_2(\varepsilon_1^2 + \bar{\varepsilon}_1^1)\varepsilon_2^{2\cdot 1} + \bar{\varepsilon}_1^1 \bar{\varepsilon}_2^2]\bar{\varepsilon}_3^3 / |\mathbf{D}|$$

$$= \left[\frac{[\hat{J}_3(\varepsilon_1^2 + \bar{\varepsilon}_1^1)\varepsilon_2^3]\varepsilon_3^4}{[\hat{J}_2(\varepsilon_1^2 + \bar{\varepsilon}_1^1)\varepsilon_2^{2\cdot 1} + \bar{\varepsilon}_1^1 \bar{\varepsilon}_2^2]\bar{\varepsilon}_3^3} + \frac{\varepsilon_3^4 + \bar{\varepsilon}_3^3}{\bar{\varepsilon}_3^3} \right]^{-1} \tag{15.25}$$

Equations (15.21)–(15.25) demonstrate the same principle as those for the control of flux through branch 2. If the enzyme at the branch point which leads to the branch associated with flux 3 were saturated by the branch point metabolite, none of the enzymes prior to the branch point, nor any of the enzymes in the competing branch, would exert any effect on flux 3 *unless* one of the metabolites in those portions of the pathway functions as an allosteric modifier of an enzyme in the portion of the pathway associated with flux 3. On the other hand, if the competing branch point enzyme, enzyme 2.1 in this case, were saturated by the branch point metabolite, the control of flux 3 exerted by those enzymes prior to the branch point would be increased while the control of those enzymes on the main pathway 'downstream', namely enzymes 3 and 4, is decreased.

It should be further noted that, as in the case of flux 2, if an enzyme prior to the branch point is insensitive to product inhibition, no enzyme 'downstream' from the insensitive enzyme but 'upstream' from the branch point will exert control of flux 3. In like manner, if an enzyme on either branch 'downstream' from the branch point were insensitive to product inhibition, the enzymes 'upstream' from the branch point and the enzymes in the other branch would be only slightly affected while the enzymes 'downstream' from the insensitive enzyme would exert no control on the flux through the branch involved. The significance of these observations is that the pathway shown in Figure 15.1 can be viewed as four relatively independent components. These are the enzymes associated with the three fluxes and the degree to which the competing enzymes are saturated by the branch point metabolite. These observations follow rather directly from the equations for the flux control coefficients, but they are not so apparent intuitively. This emphasizes the value of deriving the expressions for the control coefficients to a proper understanding of the control of a branched pathway.

The concentration control coefficients for the branched pathway in Figure 15.1 are presented in the appendix to this chapter. The reader may find it useful to derive these control coefficients.

15.2 Flux control in substrate cycles

There are numerous examples of substrate cycles in metabolism. At one time these were termed as futile cycles because it was thought that their only function was to dissipate the free energy of the hydrolysis of adenosine triphosphate (ATP) as heat. While it is possible that these cycles do serve to dissipate chemical energy as heat under certain conditions, it has become obvious that they serve a function of amplifying flux control in one direction or the other[4]. An example of such a cycle is the combination of 6-phosphofructokinase, which catalyzes the conversion of fructose-6-phosphate and ATP to form fructose diphosphate and ADP, with fructose diphosphatase, which catalyzes the hydrolysis of fructose diphosphate to

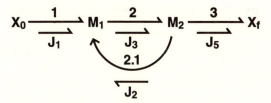

Fig. 15.2. Model of a substrate cycle.

fructose-6-phosphate and inorganic phosphate. The futile aspect of this cycle is

$$ATP + H_2O \rightarrow ADP + inorganic\ phosphate$$

This cycle is portrayed in Figure 15.2. With reference to the previous example, X_0 is glucose-6-phosphate, M_1 is fructose-6-phosphate, M_2 is fructose diphosphate and X_f would be dihydroxyacetone phosphate and glyceraldehyde-3-phosphate. The convention which has been adopted arbitrarily is that the fluxes along the main pathway are numbered with uneven numbers and the fluxes associated with branches which converge or diverage from the main pathways are numbered with even numbers. In this case, branch 2 both converges on metabolite 1 and diverges from metabolite 2. It can be seen that the following relationships exist between the fluxes

$$J_3 = J_1 + J_2 = J_2 + J_5 \quad \text{and} \quad J_1 = J_3 - J_2 = J_5.$$

The equations obtained from the sensitivity theory are

$$C_1^1 = \hat{J}_5 C_1^5 = -\bar{\varepsilon}_1^1 C_1^{M_1} + \pi_1^1 \tag{15.26}$$

$$C_2^3 = \hat{J}_2 C_2^2 + \hat{J}_5 C_2^5 = \varepsilon_1^2 C_2^{M_1} - \bar{\varepsilon}_2^2 C_2^{M_2} + \pi_2^2 \tag{15.27}$$

$$C_{2.1}^2 = -\bar{\varepsilon}_1^{2.1} C_{2.1}^{M_1} + \varepsilon_2^{2.1} C_{2.1}^{M_2} + \pi_{2.1}^{2.1} \tag{15.28}$$

$$C_3^5 = \varepsilon_2^3 C_3^{M_2} + \pi_3^3 \tag{15.29}$$

These equations give rise to the following matrix equation.

$$
\begin{vmatrix}
0 & \hat{J}_5 & \bar{\varepsilon}_1^1 & 0 \\
\hat{J}_2 & \hat{J}_5 & -\varepsilon_1^2 & \bar{\varepsilon}_2^2 \\
1 & 0 & \bar{\varepsilon}_1^{2.1} & -\varepsilon_2^{2.1} \\
0 & 1 & 0 & -\varepsilon_2^3
\end{vmatrix}
\begin{vmatrix}
C_1^2 & C_2^2 & C_{2.1}^2 & C_3^2 \\
C_1^5 & C_2^5 & C_{2.1}^5 & C_3^5 \\
C_1^{M_1} & C_2^{M_1} & C_{2.1}^{M_1} & C_3^{M_1} \\
C_1^{M_2} & C_2^{M_2} & C_{2.1}^{M_2} & C_3^{M_2}
\end{vmatrix}
=
\begin{vmatrix}
1 & 0 & 0 & 0 \\
0 & 1 & 0 & 0 \\
0 & 0 & 1 & 0 \\
0 & 0 & 0 & 1
\end{vmatrix}
$$

The denominator determinant for the control coefficients for the pathway portrayed in Figure 15.2 is

$$|D| = \hat{J}_2\hat{J}_5\bar{\varepsilon}_1^{2.1}\varepsilon_2^3 + \hat{J}_2\bar{\varepsilon}_1^1\varepsilon_2^{2.1} + \hat{J}_5(\varepsilon_1^2 + \bar{\varepsilon}_1^1)\varepsilon_2^3 + \bar{\varepsilon}_1^1\varepsilon_2^2 \tag{15.30}$$

The flux control coefficients for this pathway are

$$C_1^2 = -[\hat{J}_5\bar{\varepsilon}_1^{2.1}\varepsilon_2^3 + \bar{\varepsilon}_1^{2.1}\bar{\varepsilon}_2^2 - \varepsilon_1^2\varepsilon_1^{2.1}]/|D| \tag{15.31}$$

$$C_2^2 = [\hat{J}_5\bar{\varepsilon}_1^{2.1}\varepsilon_2^3 + \bar{\varepsilon}_1^1\,\varepsilon_1^{2.1}]/|D| \tag{15.32}$$

$$C_{2.1}^2 = [\hat{J}_5(\varepsilon_1^2 + \bar{\varepsilon}_1^1)\varepsilon_2^3 + \bar{\varepsilon}_1^1\bar{\varepsilon}_2^2]/|D| \tag{15.33}$$

$$C_3^2 = -\hat{J}_5[(\varepsilon_1^2 + \bar{\varepsilon}_1^1)\varepsilon_2^{2.1} - \bar{\varepsilon}_1^{2.1}\bar{\varepsilon}_2^2]/|D| \tag{15.34}$$

$$C_1^5 = [\hat{J}_2\bar{\varepsilon}_1^{2.1} + \varepsilon_1^2\varepsilon_2^3]/|D| \tag{15.35}$$

$$C_2^5 = \bar{\varepsilon}_1^1\varepsilon_2^3/|D| \tag{15.36}$$

$$C_{2.1}^5 = -\hat{J}_2\bar{\varepsilon}_1^1\varepsilon_2^3/|D| \tag{15.37}$$

$$C_3^5 = [\hat{J}_2\bar{\varepsilon}_1^1\varepsilon_2^{2.1} + \bar{\varepsilon}_1^1\bar{\varepsilon}_2^2]/|D| \tag{15.38}$$

The summation theorems for the pathway in Figure 15.2 are

$$\hat{J}_2C_2^2 + C_{2.1}^2 = 1 \tag{15.39}$$

$$\hat{J}_5C_1^2 + \hat{J}_5C_2^2 + C_3^2 = 0 \tag{15.40}$$

$$\hat{J}_2C_2^5 + C_{2.1}^5 = 0 \tag{15.41}$$

$$\hat{J}_5C_1^5 + \hat{J}_5C_2^5 + C_3^5 = 1 \tag{15.42}$$

The condition essential for the amplification of flux 5 is presented clearly in eq. (15.41), and with this equation and reference to eqs. (15.36) and (15.37) it can be seen that amplification will not be observed if flux 2 were equal to zero or if enzyme 1 were completely insensitive to product inhibition or if enzyme 3 were saturated by metabolite 2. If either of the latter two conditions were true, neither enzyme 2 nor enzyme 2.1 would exert control on flux 5. However, if none of the three conditions-mentioned before were true, the control exerted by enzyme 2 on flux 5 would be amplified. This amplification is basically due to the effect of branching in the pathway. In the case of a substrate cycle, the degree of amplification is dependent on the flux in the opposite direction of the main pathway. It is significant that two of the enzymes which have been reported to play a significant role in the control of glycolysis are 6-phosphofructokinase and pyruvate kinase, and both of these enzymes are constituent enzymes in substrate cycles[4].

The foregoing considerations provide a quantitative basis for the amplification of control by enzymes involved in substrate cycles, although this metabolic role of substrate cycles was not obvious by intuition. Another feature which has been demonstrated clearly in this chapter is that the degree of saturation of branch point enzymes and the sensitivity of enzymes about the branch point play an important role in the metering of flux through competing pathways in metabolism.

15.3 Problems for chapter 15

15.1 Figure 15.3 presents a multi-enzyme pathway. Using the procedures described in this and the previous chapters, obtain the matrices necessary for derivation of the control coefficients for the foregoing pathway.

15.2 Multiply the A and B matrices of problem 15.1 to obtain the relationships, the flux control coefficients and the concentration control coefficients and then reverse the order of the multiplication of matrices A and B to obtain the summation and connectivity theorems of MCT.

15.3 Derive the expressions for the flux control coefficients for both J_2 and J_3 and the concentration control coefficients for the pathway portrayed in problem 15.1.

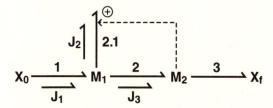

Fig. 15.3. Model of a multi-enzyme system with a single divergent branch where a metabolite in one branch activates the competing branch.

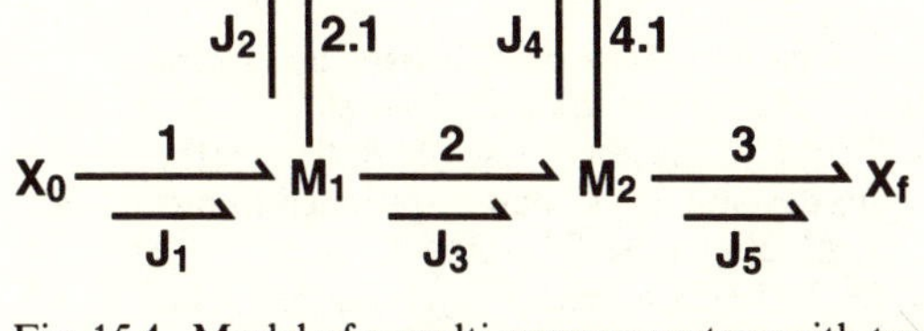

Fig. 15.4. Model of a multi-enzyme system with two divergent branches.

15.4   Figure 15.4 multi-enzyme presents a pathway. Derive the expressions for presents a flux control coefficients for J_2, J_4 and J_5 and the
concentration control coefficients for the foregoing multi-enzyme
system.

Appendix

The following are the concentration control coefficients for the pathway portrayed
in Fig. 15.1. The denominator determinant is the same as for the flux control
coefficients.

$$C_1^{M_1} = [(\hat{J}_2 \varepsilon_2^{2\cdot1} + \bar{\varepsilon}_2^2)(\varepsilon_3^3 + \bar{\varepsilon}_3^3) + \hat{J}_3 \varepsilon_2^3 \varepsilon_3^4]/|D|$$

$$C_2^{M_1} = -[\hat{J}_2 \varepsilon_2^{2\cdot1}(\varepsilon_3^4 + \bar{\varepsilon}_3^3) + \hat{J}_3 \varepsilon_2^3 \varepsilon_3^4]/|D|$$

$$C_{2.1}^{M_1} = -\hat{J}_2 \bar{\varepsilon}_2^2(\varepsilon_3^4 + \bar{\varepsilon}_3^3)/|D|$$

$$C_3^{M_1} = -\hat{J}_3 \bar{\varepsilon}_2^2 \varepsilon_3^4/|D|$$

$$C_4^{M_1} = -\hat{J}_3 \bar{\varepsilon}_2^2 \bar{\varepsilon}_3^3/|D|$$

$$C_1^{M_2} = \varepsilon_1^2(\varepsilon_3^4 + \bar{\varepsilon}_3^3)/|D|$$

$$C_2^{M_2} = \bar{\varepsilon}_1^1(\varepsilon_3^4 + \bar{\varepsilon}_3^3)/|D|$$

$$C_{2.1}^{M_2} = -\hat{J}_2(\varepsilon_1^2 + \bar{\varepsilon}_1^1)(\varepsilon_3^4 + \bar{\varepsilon}_3^3)/|D|$$

$$C_3^{M_2} = -\hat{J}_3(\varepsilon_1^2 + \bar{\varepsilon}_1^1)\varepsilon_3^4/|D|$$

$$C_4^{M_2} = -\hat{J}_3(\varepsilon_1^2 + \bar{\varepsilon}_1^1)\bar{\varepsilon}_3^3/|D|$$

$$C_1^{M_3} = \varepsilon_1^2 \varepsilon_2^3/|D|$$

$$C_2^{M_3} = \bar{\varepsilon}_1^1 \varepsilon_2^3/|D|$$

$$C_{2.1}^{M_3} = -\hat{J}_2(\varepsilon_1^2 + \bar{\varepsilon}_1^1)\varepsilon_2^3/|D|$$

$$C_3^{M_3} = [\hat{J}_2(\varepsilon_1^2 + \bar{\varepsilon}_1^1) + \bar{\varepsilon}_1^1 \bar{\varepsilon}_2^2]/|D|$$

$$C_4^{M_3} = -[\hat{J} \varepsilon_2^{2\cdot1} + \hat{J}_3 \varepsilon_2^3](\varepsilon_1^2 + \bar{\varepsilon}_1^1) + \bar{\varepsilon}_1^1 \bar{\varepsilon}_2^2/|D|$$

References

1. La Porte, D. C., Walsh, K. and Koshland, D. E. (1984). The branch point
    effect. Ultrasensitivity and subsensitivity to metabolic control. *J. Biol. Chem.*
    259: 14068–75.
2. Holzer, H. (1961). Regulation of carbohydrate metabolism by enzyme
    competition. *Cold Springs Harb. Symp. Quant. Biol.* **26**: 277–88.
3. Cascante, M., Franco, R. and Canela, E. I. (1989). Use of implicit methods
    from general sensitivity theory to develop a systematic approach to metabolic
    control. II. Complex systems. *Math. Biosci.* **94**: 289–309.
4. Newsholme, E. A. and Crabtree, B. (1976). Substrate cycles in metabolic
    regulation and in heat generation. *Biochem. Soc. Symp.* **41**: 61–109.

16
Biochemical systems theory

An approach to the mathematical analysis of multi-enzyme systems will be presented in this chapter which may appear distinct from that presented in chapters 14 and 15 but which is actually related to the previous treatment in a fundamental manner. The biochemical systems theory (BST) has been developed through the initiative of M. A. Savageau[1-4]. Biochemical systems theory is a more general treatment of metabolic control than the metabolic control theory (MCT) which was presented in the previous chapters[5,6].

16.1 Power law formulation of control of a linear multi-enzyme system

The observation that the rate of most biological reactions can be described by rational polynomial equations is basic to BST. A general form of a rational polynomial is

$$v = \frac{\alpha_n X^{n-1} + \alpha_{n-1} X^{n-1} + \cdots + \alpha_1 X + \alpha_0}{\beta_n X^n + \beta_{n-1} X^{n-1} + \cdots + \beta_1 X + \beta_0} \tag{16.1}$$

Equation (16.1) can be written in a factored form.

$$v = \frac{\Phi(a_n + X)(a_{n-1} + X) \cdots (a_1 + X)}{\theta(b_n + X)(B_{n-1} + X) \cdots (b_1 + X)} \tag{16.2}$$

The expression of eq. (16.2) in logarithmic form gives

$$\ln v = \ln \frac{\Phi}{\theta} + \ln(a_n + X) + \ln(a_{n-1} + X) + \cdots + \ln(a_1 + X)$$
$$- \ln(b_n + X) - \ln(b_{n-1} + X) - \cdots - \ln(b_1 + X) \tag{16.3}$$

In view of eq. (16.3), it is not surprising that rational polynomials give rise to plots which consist of linear segments when they are plotted in log-log space. This is true when the rates of most biological reactions are plotted in log-log space against the concentration of the independent variable. In fact, in many instances, the relationship is linear over a number of orders of magnitude of the independent variable[3]. This observation led Savageau to develop a power law formalism based on Taylor's theorem when plotted in logarithmic space.

Taylor's theorem is one of a number of mean value theorems which allow one to relate the behavior of any continuous function with its derivatives in a given interval. This, of course, necessitates that the function be continuous and differentiable within the given interval. Taylor's theorem (series) is one of the most useful and powerful of the mean value theorems. A full derivation of Taylor's theorem is beyond the scope of this text, but for the reader who is unfamiliar with the concept behind the mean value theorems, the following brief discussion is included.

The first derivative of any function which is continuous and differentiable in the interval (a, b) is given by eq. (16.4).

$$f_{(\zeta)}^1 = \frac{f_{(b)} - f_{(a)}}{b - a}, \quad \text{where} \quad a < \zeta < b. \tag{16.4}$$

If the interval (a, b) is small, eq. (16.4) approximately becomes

$$f_{(b)} \cong f_{(a)} + f_{(a)}^1(b - a) \tag{16.5}$$

The reader should recall that this procedure was employed in chapter 12 to obtain the point of intersection of the y axis by an asymptote. The actual expression for a Taylor series is

$$f_{(b)} = f_{(a)} + f_{(a)}^1(b - a) + f_{(a)}^2 \frac{(b - a)^2}{2!} + \cdots + f_{(a)}^n \frac{(b - a)^n}{n!} + R \tag{16.6}$$

where R is the remainder. If the function is linear in the interval (a, b), the second and higher derivatives are equal to zero, and the Taylor series can be truncated to the first two terms on the right-hand side of eq. (16.6) plus the remainder.

If the truncated Taylor series is applied in logarithmic space to an enzymic reactions whose rate is v_j and whose substrate is M_i the following equation is obtained:

$$\ln v_j = \ln v_{j_0} + \left(\frac{d \ln v_j}{d \ln M_i} \right) (\ln M_i - \ln M_{i_0}) \tag{16.7}$$

In eq. (16.7), the subscripts i_0 and j_0 refer to M_i and v_j about a specific operating point. From the definition for elasticity given in chapter 14,

$\varepsilon_i^j = (d \ln v_j)/(d \ln M_i)$ where the differential is evaluated at i_0 and j_0. In chapter 14, the elasticity coefficient was defined as the fractional change in velocity brought about by a fractional change in metabolite concentration as a matter of rendering the term "change" non-dimensional. Equation (16.7) provides an explicit explanation for this definition. If the following definition is made

$$\ln \alpha_j = \ln v_{j_0} + \varepsilon_i^j - \varepsilon_i^j \ln M_{i_0} \tag{16.8}$$

Substitution of eq. (16.8) into eq. (16.7) gives

$$\ln v_j = \ln \alpha_j + \varepsilon_i^j \ln M_i \tag{16.9}$$

$$v_j = \alpha_j M_i^{\varepsilon_i^j} \tag{6.10}$$

In BST terminology, α_j is called a rate constant. However, if more than one metabolite in the pathway affects the rate of the reaction eq. (16.10) becomes

$$v_j = \alpha_j \prod_{k=1}^{n} M_i^{\varepsilon_{ik}^j} \tag{16.11}$$

Consider once again the linear multi-enzyme pathway shown in Figure 14.1. The intermediate metabolites are assumed to be in steady state and therefore if one considers any one of these metabolites

$$\frac{dM_i}{dt} = v_i - v_j = 0 \tag{16.12}$$

If β_j is employed as the symbol for the rate constant for the reaction(s) which remove rather than synthesize the metabolite, and $\bar{\varepsilon}_i^i = -\varepsilon_i^i$; substitution of eq. (16.11) into eq. (16.12) results in the following differential equations for the intermediate metabolites for the pathway in Figure 16.1.

$$\frac{dM_1}{dt} = \alpha_1 X_0^{\varepsilon_0^1} M_1^{-\bar{\varepsilon}_1^1} - \beta_1 M_0^{\varepsilon_0^2} M_2^{-\bar{\varepsilon}_2^2} = 0 \tag{16.13}$$

$$\frac{dM_2}{dt} = \alpha_2 M_1^{\varepsilon_1^2} M_2^{-\bar{\varepsilon}_2^2} - \beta_2 M_2^{\varepsilon_2^3} M^{-\bar{\varepsilon}_3^3} = 0 \tag{16.14}$$

$$\frac{dM_3}{dt} = \alpha_3 M_2^{\varepsilon_2^3} M_3^{-\bar{\varepsilon}_3^3} - \beta_3 M_3^{\varepsilon_3^4} = 0 \tag{16.15}$$

$$X_0 \xrightarrow{1} M_1 \xrightarrow{2} M_2 \xrightarrow{3} M_3 \xrightarrow{4} X_f$$

Fig. 16.1. A linear multi-enzyme system.

These foregoing equations can be rearranged as the following:

$$b_1 = \varepsilon_0^1 y_0 - (\varepsilon_1^2 + \bar{\varepsilon}_1^1) y_1 + \bar{\varepsilon}_2^2 y_2 \tag{16.16}$$

$$b_2 = \varepsilon_1^2 y_1 - (\varepsilon_2^3 + \bar{\varepsilon}_2^2) y_2 + \bar{\varepsilon}_3^3 y_3 \tag{16.17}$$

$$b_3 = \varepsilon_2^3 y_2 - (\varepsilon_3^4 + \bar{\varepsilon}_3^3) y_3 \tag{16.18}$$

where $b_i = \ln(\beta_i/\alpha_i)$, $y_0 = \ln X_0$, and $y_1 = \ln M_i$. These equations can be expressed in matrix form.

$$\begin{vmatrix} -(\varepsilon_1^2 + \bar{\varepsilon}_1^1) & \bar{\varepsilon}_2^2 & 0 \\ \varepsilon_1^2 & -(\varepsilon_2^3 + \bar{\varepsilon}_2^2) & \bar{\varepsilon}_3^3 \\ 0 & \varepsilon_2^3 & -(\varepsilon_3^4 + \bar{\varepsilon}_3^4) \end{vmatrix} \begin{vmatrix} y_1 \\ y_2 \\ y_3 \end{vmatrix} = \begin{vmatrix} b - \varepsilon_0^1 y_0 \\ b_2 \\ b_3 \end{vmatrix}$$

The solution of this system of equations provides a means by which it is possible to estimate the logarithmic concentration of the intermediate metabolites in a metabolic pathway. The denominator determinant of these expressions is

$$|\mathbf{D}| = -\left[\varepsilon_1^2 \varepsilon_2^3 \varepsilon_3^4 + \bar{\varepsilon}_1^1 \varepsilon_2^3 \varepsilon_3^4 + \bar{\varepsilon}_1^1 \bar{\varepsilon}_2^2 \varepsilon_3^4 + \bar{\varepsilon}_1^1 \bar{\varepsilon}_2^2 \bar{\varepsilon}_3^3 \right] \tag{16.19}$$

It should be noted that eq. (16.19) is identical to the denominator determinant for the control coefficients for the mechanism portrayed in both Figure 14.1 and Figure 16.1. The following are the expressions for the logarithmic concentrations:

$$y_1 = \left[\varepsilon_0^1 (\varepsilon_2^3 \varepsilon_3^4 + \bar{\varepsilon}_2^2 \varepsilon_3^4 + \bar{\varepsilon}_2^2 \bar{\varepsilon}_3^3) y_0 - (\varepsilon_2^3 \varepsilon_3^4 + \bar{\varepsilon}_2^2 \varepsilon_3^4 + \bar{\varepsilon}_2^2 \bar{\varepsilon}_3^3) b_1 \right.$$

$$\left. - \bar{\varepsilon}_2^2 (\varepsilon_3^4 + \bar{\varepsilon}_3^3) b_2 - \bar{\varepsilon}_2^2 \bar{\varepsilon}_3^3 b_3 \right] / |\mathbf{D}| \tag{16.20}$$

$$y_2 = \left[\varepsilon_0^1 \varepsilon_1^2 (\varepsilon_3^4 + \bar{\varepsilon}_3^3) y_0 - \varepsilon_1^2 (\varepsilon_3^4 + \bar{\varepsilon}_3^3) b_1 - (\varepsilon_1^2 + \bar{\varepsilon}_1^1)(\varepsilon_3^4 + \bar{\varepsilon}_3^3) b_2 \right.$$

$$\left. - (\varepsilon_1^2 + \bar{\varepsilon}_1^1) \varepsilon_3^3 b_3 \right] / |\mathbf{D}| \tag{16.21}$$

$$y_3 = \left[\varepsilon_0^1 \varepsilon_1^2 \varepsilon_2^3 y_0 - \varepsilon_1^2 \varepsilon_2^3 b_1 - (\varepsilon_1^2 + \bar{\varepsilon}_1^1) \varepsilon_2^3 b_2 \right.$$

$$\left. - (\varepsilon_1^2 \varepsilon_2^3 + \bar{\varepsilon}_1^1 \varepsilon_2^3 + \bar{\varepsilon}_1^1 \bar{\varepsilon}_2^2) b_3 \right] / [\mathbf{D}] \tag{16.22}$$

The sign of the Y values are all positive because the numerators of the expressions are negative as is the denominator. The ability to obtain an estimate of the concentrations of the intermediate metabolites is a major advantage. Among other practical benefits, this makes it possible to conduct simulations such that the dynamic stability of the system can be investigated[3,7]. It was believed previously, that MCT did not provide the opportunity to estimate the concentration of intermediate metabolites.

However, consider the expressions for the concentration control coefficients for the pathway in Figure 16.1, which were presented in chapter 14.

$$C_1^{M_1} = (\varepsilon_2^3 \varepsilon_3^4 + \bar{\varepsilon}_2^2 \varepsilon_3^4 + \bar{\varepsilon}_2^2 \bar{\varepsilon}_3^3)/|D| \tag{16.23}$$

$$C_2^{M_1} = -\varepsilon_2^3 \varepsilon_3^4/|D| \tag{16.24}$$

$$C_3^{M_1} = -\bar{\varepsilon}_2^2 \varepsilon_3^4/|D| \tag{16.25}$$

$$C_4^{M_1} = -\bar{\varepsilon}_2^2 \bar{\varepsilon}_3^3/|D| \tag{16.26}$$

$$C_1^{M_2} = \varepsilon_1^2 (\varepsilon_3^4 + \bar{\varepsilon}_3^3)/|D| \tag{16.27}$$

$$C_2^{M_2} = \bar{\varepsilon}_1^1 (\varepsilon_3^4 + \bar{\varepsilon}_3^3)/|D| \tag{16.28}$$

$$C_3^{M_2} = -(\varepsilon_1^2 + \bar{\varepsilon}_1^1)\varepsilon_3^4/|D| \tag{16.29}$$

$$C_4^{M_2} = -(\varepsilon_1^2 + \bar{\varepsilon}_1^1)\bar{\varepsilon}_3^3/|D| \tag{16.30}$$

$$C_1^{M_3} = \varepsilon_1^2 \varepsilon_2^3/|D| \tag{16.31}$$

$$C_2^{M_3} = \bar{\varepsilon}_1^1 \varepsilon_2^3/|D| \tag{16.32}$$

$$C_3^{M_3} = \bar{\varepsilon}_1^1 \bar{\varepsilon}_2^2/|D| \tag{16.33}$$

$$C_4^{M_3} = -(\varepsilon_1^2 \varepsilon_2^3 + \bar{\varepsilon}_1^1 \varepsilon_2^3 + \bar{\varepsilon}_1^1 \bar{\varepsilon}_2^2)/|D| \tag{16.34}$$

When eq. (16.20) is compared with the expressions for the concentration control coefficients for M_1 it is apparent that the equation for the logarithmic concentration of M_1 can be expressed in terms of the concentration control coefficients. That is, eq. (16.20) can be written as

$$y_1 = \varepsilon_0^1 C_1^{M_1} y_0 + (C_2^{M_1} + C_3^{M_1} + C_4^{M_1})b_1 + (C_3^{M_1} + C_4^{M_1})b_2 + C_4^{M_1}b_3 \tag{16.35}$$

In like manner, the expressions for y_2 and y_3 can be written as

$$y_2 = \varepsilon_0^1 C_1^{M_2} y_0 + (C_2^{M_2} + C_3^{M_2} + C_4^{M_2})b_1 + (C_3^{M_2} + C_4^{M_2})b_2 + C_4^{M_2}b_3 \tag{16.36}$$

$$y_3 = \varepsilon_0^1 C_1^{M_3} y_0 + (C_2^{M_3} + C_3^{M_3} + C_4^{M_3})b_1 + (C_3^{M_3} + C_4^{M_3})b_2 + C_4^{M_3}b_3 \tag{16.37}$$

Equations (16.35)–(16.37) demonstrate a relationship between BST and MCT which was not recognized previously. For linear pathways, if expressions have been obtained for the concentration control coefficients, the equations for the logarithmic concentrations can be written without going through the procedure outlined earlier in this chapter. The converse is also true. A general equation for the logarithmic concentration of the intermediate metabolites in a linear pathway is

$$\left[y_i = \varepsilon_0^1 C_1^{M_i} y_0 + \sum_{j=1}^{n} \sum_{k=j+1}^{n} C_k^{M_i} b_j \right]_{i=1,2,\ldots m} \tag{16.38}$$

where m is the number of metabolites and n is the number of enzymes.

It must be pointed out that the variant of BST which has been presented in this text is the generalized mass action (GMA) variant. Savageau et al. prefer the S-system variant of BST[4,8]. The differences between these variants is in the level of aggregation. Aggregation is effected at the level of the individual reactions in the GMA variant. That is the procedure which has been followed in this chapter. In contrast, in the S-system all the reactions which result in synthesis of the metabolite are aggregated together and all the reactions which result in removal of the metabolite are aggregated together. It is acknowledged that Voit and Savageau have demonstrated some advantages to the S-system[8], but the GMA variant is not invalid. This author prefers the GMA variant because the S-system tends to conceal some of the synergism which can take place between the components of the pathway.

The term *metabolic control* conjures thoughts of the control of flux through a pathway. The control of flux is, indeed, an important consideration in metabolic control. However, Savageau has emphasized that minimization of intermediate metabolites is, no doubt, one driving force in the process of evolution[3]. Others have emphasized the burden placed on the cell by the necessity to solubilize metabolites and other cellular components[9,10]. There are many implications of this facet of metabolic control. The loss of the ability to synthesize the indispensable amino acids by higher animals is, no doubt, a means of reducing the number and magnitude of metabolic pools. Metabolite channelling is another means of reducing the size of metabolite pools[11]. Many other examples of this concept might be cited. For this reason, it is informative to differentiate eqs. (16.35)–(16.37) with respect to y_0, for this provides a quantitative measure of the effect of the independent variable on the concentration of the intermediate metabolites. This is called the logarithmic gain, and it is defined in eq. (16.39).

$$L_{io} = \frac{\partial y_i}{\partial y_0} = \varepsilon_0^1 C_1^{M_i}, \quad i = 1, 2, \ldots, m \tag{16.39}$$

The sensitivity of the intermediate metabolites to the rate constants can be obtained by differentiating eq. (16.35)–(16.37) with respect to the b_i values.

$$\left[S(M_i, b_j) = \frac{\partial y_i}{\partial b_j} = \sum_{k=j+1}^{n} C_k^{M_i} \right]_{i=1,2,\ldots m} \tag{16.40}$$

16.2 Power law formulation of control of linear multi-enzyme pathways when enzyme activities are variable

One of the advantages of BST is that it provides for the treatment of enzyme activities as variables rather than as parameters of the system. This can be demonstrated with the pathway portrayed in Figure 16.1. If all the enzyme activities are taken as variables, the differential equations for metabolites are as follows:

$$\frac{dM_1}{dt} = \alpha_1 X_0^{\varepsilon_0^1} M_1^{-\bar{\varepsilon}_1^1} E_1^{\pi_1^1} - \beta_1 M_1^{\varepsilon_1^2} M_2^{-\bar{\varepsilon}_2^2} E_2^{\pi_2^2} = 0 \tag{16.41}$$

$$\frac{dM_2}{dt} = \alpha_2 M_1^{\varepsilon_1^2} M_2^{-\bar{\varepsilon}_2^2} E_2^{\pi_2^2} - \beta_2 M_2^{\varepsilon_2^3} M_3^{-\bar{\varepsilon}_3^3} E_3^{\pi_3^3} = 0 \tag{16.42}$$

$$\frac{dM_3}{dt} = \alpha_3 M_2^{\varepsilon_2^3} M_3^{-\bar{\varepsilon}_3^3} E_3^{\pi_3^3} - \beta_3 M_3^{\varepsilon_3^4} E_4^{\pi_4^4} = 0 \tag{16.43}$$

Once again, the reader should be aware that, for the sake of consistency, the symbolism employed in this text is more consistent with MCT than BST. Equations (16.41)–(16.43) can be expressed in logarithmic form.

$$b_1 - \varepsilon_0^1 y_0 + \pi_1^1 y_{e_1} + \pi_2^2 y_{e_2} = -(\varepsilon_1^2 + \bar{\varepsilon}_1^1)y_1 + \bar{\varepsilon}_2^2 y_2 \tag{16.44}$$

$$b_2 - \pi_2^2 y_{e_2} + \pi_3^3 y_{e_3} = \varepsilon_1^2 y_1 - (\varepsilon_2^3 + \bar{\varepsilon}_2^2)y_2 + \bar{\varepsilon}_3^3 y_3 \tag{16.45}$$

$$b_3 - \pi_3^3 y_{e_3} + \pi_4^4 y_{e_4} = \varepsilon_2^3 y_2 - (\varepsilon_3^4 + \bar{\varepsilon}_3^3)y_3 \tag{16.46}$$

In the foregoing equations, $y_{e_i} = \ln E_i$. These expressions give rise to the following matrices:

$$\begin{vmatrix} -(\varepsilon_1^2 + \bar{\varepsilon}_1^1) & \bar{\varepsilon}_2^2 & 0 \\ \varepsilon_1^2 & -(\varepsilon_2^3 + \bar{\varepsilon}_2^2) & \bar{\varepsilon}_3^3 \\ 0 & \varepsilon_2^3 & -(\varepsilon_3^4 + \bar{\varepsilon}_3^3) \end{vmatrix} \begin{vmatrix} y_i \\ y_2 \\ y_3 \end{vmatrix}$$

$$= \begin{vmatrix} -\varepsilon_0^1 & -\pi_1^1 & \pi_2^2 & 0 & 0 \\ 0 & 0 & -\pi_2^2 & \pi_3^3 & 0 \\ 0 & 0 & 0 & -\pi_3^3 & \pi_4^4 \end{vmatrix} \begin{vmatrix} y_0 \\ y_{e_1} \\ y_{e_2} \\ y_{e_3} \\ y_{e_4} \end{vmatrix} + \begin{vmatrix} b_1 \\ b_2 \\ b_3 \end{vmatrix}$$

The foregoing matrices can be abbreviated as follows:

$$|A||y|_m = |B||y|_i + |b| \tag{16.47}$$

where $|y|_m$ is a vector of logarithms of metabolites and $|y|_i$ is a vector of logarithms of independent variables. A more convenient form of eq. (16.47) is

$$|y|_m = |A|^{-1}|B||y|_i + |A|^{-1}|b| \tag{16.48}$$

where $|A|^{-1}$ is the matrix whose elements are the inverse of $|A|$. The inverse of $|A|$ can be defined as $|A|^{-1} = (|M|/|D|)$ where $|M|$ is the adjoint of $|A|$, and $|D|$ is the symbolic determinant of A. The useful concept of the adjoint of a matrix is presented in the appendix of this chapter. The previous system of matrices can be written as

$$\begin{vmatrix} y_1 \\ y_2 \\ y_3 \end{vmatrix} = \frac{1}{|D|} \begin{vmatrix} M_{11} & M_{12} & M_{13} \\ M_{21} & M_{22} & M_{23} \\ M_{31} & M_{32} & M_{33} \end{vmatrix} \begin{vmatrix} -\varepsilon_1^0 & -\pi_1^1 & \pi_2^2 & 0 & 0 \\ 0 & 0 & -\pi_2^2 & \pi_3^3 & 0 \\ 0 & 0 & 0 & -\pi_3^3 & \pi_4^4 \end{vmatrix} \begin{vmatrix} y_0 \\ y_{e_1} \\ y_{e_2} \\ y_{e_3} \\ y_{e_4} \end{vmatrix}$$

$$+ \frac{1}{|D|} \begin{vmatrix} M_{11} & M_{12} & M_{13} \\ M_{21} & M_{22} & M_{23} \\ M_{31} & M_{32} & M_{33} \end{vmatrix} \begin{vmatrix} b_1' \\ b_2' \\ b_3' \end{vmatrix}$$

The equations for the logarithmic concentration of the intermediate metabolites are as follows:

$$y_1 = [-\varepsilon_1^0 M_{11} y_0 - \pi_1^1 M_{11} y_{e_1} + \pi_2^2 (M_{11} - M_{12}) y_{e_2} + \pi_3^3 (M_{12} - M_{13}) y_{e_3}$$
$$+ \pi_4^4 M_{13} y_{e_4} + M_{11} b_1 + M_{12} b_2 + M_{13} b_3]/|D| \tag{16.49}$$

$$y_2 = [-\varepsilon_0^1 M_{21} y_0 - \pi_1^1 M_{21} y_{e_1} + \pi_2^2 (M_{21} - M_{23}) y_{e_2} + \pi_3^3 (M_{22} - M_{23}) y_{e_3}$$
$$+ \pi_4^4 M_{23} y_{e_4} + M_{21} b_1 + M_{22} b_2 + M_{23} b_3]/|D| \tag{16.50}$$

$$y_3 = [-\varepsilon_0^1 M_{31} y_0 - \pi_1^1 M_{31} y_{e_1} + \pi_2^2 (M_{31} - M_{32}) y_{e_2} + \pi_3^3 (M_{32} - M_{33}) y_{e_3}$$
$$+ \pi_4^4 M_{33} y_{e_3} + M_{31} b_1 + M_{32} b_2 + M_{33} b_3]/|D| \tag{16.51}$$

Equations (16.49)–(16.51) are appropriate within the context of the GMA variant of BST, but they apply to the treatment of MCT presented in this text *only* if $\pi_1^1 = \pi_2^2 = \pi_3^3 = \pi_4^4 = 1$ and if there are no interactions between enzymes. However, later expansions of MCT provide for enzyme–enzyme interactions[12].

Inversion of matrix A provides the following values of the elements in matrix M

$$M_{11}/|D| = -C_1^{M_1}, \quad M_{12}/|D| = C_3^{M_1} + C_4^{M_1}, \quad M_{13}/|D| = C_4^{M_1},$$

$$M_{21}/|D| = -C_1^{M_2}, \quad M_{22}/|D| = C_3^{M_2} + C_4^{M_2}, \quad M_{23}/|D| = C_4^{M_2},$$

$$M_{31}/|D| = -C_1^{M_3}, \quad M_{32}/|D| = C_3^{M_3} + C_4^{M_3}, \quad M_{33}/|D| = C_4^{M_3}.$$

The equations for the logarithmic concentration of the intermediate metabolites under the conditions where MCT treatment presented in this text are valid, are

$$y_1 = \varepsilon_0^1 C_1^{M_1} y_0 + C_1^{M_1} y_{e_1} + C_2^{M_1} y_{e_2} + C_3^{M_1} y_{e_3} + C_4^{M_1} y_{e_4}$$
$$+ (C_2^{M_1} + C_3^{M_1} + C_4^{M_1}) b_1' + (C_3^{M_1} + C_4^{M_1}) b_2' + C_4^{M_1} b_3' \tag{16.52}$$

$$y_2 = \varepsilon_0^1 C_1^{M_2} y_0 + C_1^{M_2} y_{e_1} + C_2^{M_2} y_{e_2} + C_3^{M_2} y_{e_3} + C_4^{M_2} y_{e_4}$$
$$+ (C_2^{M_2} + C_3^{M_2} + C_4^{M_2}) b_1' + (C_3^{M_2} + C_4^{M_2}) b_2' + C_4^{M_2} b_3' \tag{16.53}$$

$$y_3 = \varepsilon_0^1 C_1^{M_3} y_0 + C_1^{M_3} y_{e_3} + C_2^{M_3} y_{e_2} + C_3^{M_3} y_{e_3} + C_4^{M_3} y_{e_4}$$
$$+ (C_2^{M_3} + C_3^{M_3} + C_4^{M_3}) b_1' + (C_3^{M_3} + C_4^{M_3}) b_2' + C_4^{M_3} b_3' \tag{16.54}$$

The significant point is that, if the assumptions of the original MCT treatment are valid, then the equations for the logarithmic concentrations of the intermediate metabolites can be written in terms of ε_0^1, concentration control coefficients and rate constants in the case of linear pathways.

16.3 Power law formulation for branched pathways

The multi-enzyme pathway portrayed in Figure 15.1 will be used as the model for power law formulation of a branched pathway.

The equation derived in chapter 15 to relate the three fluxes in the pathway was

$$C_i^1 = \hat{J}_2 C_i^2 + \hat{J}_3 C_i^3 \tag{16.55}$$

where $C_i^j = C_{E_i}^{J_j}$ and $\hat{J}_i = (J_i/J_1)$. Since aggregation is performed at the level of the individual reaction in the GMA variant of BST employed in this text, the β term at the branch point is separated into a β term for each branch. This rate constant is multiplied by the concentration of the branch point metabolite raised to the power of the fractional flux flowing through the branch times the elasticity of the branch enzyme for its substrate. This

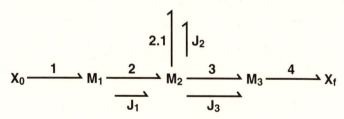

Fig. 16.2. A multi-enzyme system with one divergent branch.

procedure is applicable *provided* there are no cross links between the branches and it is illustrated by the following differential equations for the intermediate metabolites in the pathway shown in Figure 16.2

$$\frac{dM_1}{dt} = \alpha_1 X_0^{\varepsilon_0^1} M_1^{-\bar{\varepsilon}_1^1} - \beta_1 M_0^{\varepsilon_1^2} M_2^{-\bar{\varepsilon}_2^2} = 0 \tag{16.56}$$

$$\frac{dM_2}{dt} = \alpha_2 M_1^{\varepsilon_1^2} M_2^{-\bar{\varepsilon}_2^2} - (\beta_2 M_0^{J_3 \varepsilon_3^2} M_3^{-\bar{\varepsilon}_3^3}) = 0 \tag{16.57}$$

$$\frac{dM_3}{dt} = \alpha_3 M_2^{J_3 \varepsilon_2^3} M_3^{-\bar{\varepsilon}_3^3} - \beta_3 M_3^{\varepsilon_3^4} = 0 \tag{16.58}$$

The logarithmic equations derived from eqs. (16.56)–(16.58) are the following:

$$b_1 = \varepsilon_0^1 y_0 - (\varepsilon_1^2 + \bar{\varepsilon}_1^1) y_1 + \bar{\varepsilon}_2^2 y_2 \tag{16.59}$$

$$b_2 + b_{2.1} = \varepsilon_1^2 y_1 - (\bar{\varepsilon}_2^2 + \hat{J}_2 \varepsilon_2^{2.1} + \hat{J}_3 \varepsilon_2^3) y_2 + \bar{\varepsilon}_3^3 y_3 \tag{16.60}$$

$$b_3 = \hat{J}_3 \varepsilon_2^3 y_2 - (\varepsilon_3^4 + \bar{\varepsilon}_3^3) y_3 \tag{16.61}$$

The equations for the logarithmic concentrations of the metabolites are derived by analysis of the following system of matrices:

$$
\begin{vmatrix}
-(\varepsilon_1^2 + \bar{\varepsilon}_1^1) & \bar{\varepsilon}_2^2 & 0 \\
\varepsilon_1^2 & -(\bar{\varepsilon}_2^2 + \hat{J}_2 \varepsilon_2^{2.1} + \hat{J}_3 \varepsilon_2^3) & \bar{\varepsilon}_3^3 \\
0 & \hat{J}_3 \varepsilon_2^3 & -(\varepsilon_3^4 + \bar{\varepsilon}_3^3)
\end{vmatrix}
\begin{vmatrix} y_1 \\ y_2 \\ y_3 \end{vmatrix}
=
\begin{vmatrix} b_1 - \varepsilon_0^1 y_0 \\ b_2 + b_{2.1} \\ b_3 \end{vmatrix}
$$

The denominator determinant is as follows:

$$|D| = -[\hat{J}_2(\varepsilon_1^2 + \bar{\varepsilon}_1^1)\varepsilon_2^{2.1}(\varepsilon_3^4 + \bar{\varepsilon}_3^3) + \hat{J}_3(\varepsilon_1^2 + \bar{\varepsilon}_1^1)\varepsilon_2^3 \varepsilon_3^4 + \varepsilon_1^1 \bar{\varepsilon}_2^2(\varepsilon_3^4 + \bar{\varepsilon}_3^3)] \tag{16.62}$$

It should be recognized that eq. (16.62) is identical to eq. (15.15). The following are the equations for the logarithmic concentrations of the metabolites.

$$y_1 = [\varepsilon_0^1[(\hat{J}_2 \varepsilon_2^{2.1} + \bar{\varepsilon}_2^2)(\varepsilon_3^4 + \bar{\varepsilon}_3^3) + \hat{J}_3 \varepsilon_2^3 \varepsilon_3^4] y_0 - [(\hat{J}_2 \varepsilon_2^{2.1} + \bar{\varepsilon}_2^2)(\varepsilon_3^4 + \bar{\varepsilon}_3^3)$$
$$+ \hat{J}_3 \varepsilon_2^3 \varepsilon_3^4] b_1 - \bar{\varepsilon}_2^2(\varepsilon_3^4 + \bar{\varepsilon}_3^3) b_{2.1} - \bar{\varepsilon}_2^2(\varepsilon_3^4 + \bar{\varepsilon}_3^3) b_2 - \bar{\varepsilon}_2^2 \bar{\varepsilon}_3^3 b_3]/|D| \tag{16.63}$$

$$y_2 = \varepsilon_0^1 \varepsilon_1^2 (\varepsilon_3^4 + \bar{\varepsilon}_3^3) y_0 - \varepsilon_1^2(\varepsilon_3^4 + \bar{\varepsilon}_3^3) b_1 - [(\varepsilon_1^2 + \bar{\varepsilon}_1^1)(\varepsilon_3^4 + \bar{\varepsilon}_3^3) b_{2.1}$$
$$- (\varepsilon_1^2 + \bar{\varepsilon}_1^1)(\varepsilon_3^4 + \bar{\varepsilon}_3^3) b_2 - (\varepsilon_1^2 + \bar{\varepsilon}_1^1)\bar{\varepsilon}_3^3 b_3]/|D| \tag{16.64}$$

$$y_3 = \hat{J}_3 \varepsilon_0^1 \varepsilon_1^2 \varepsilon_2^3 y_0 - \hat{J}_3 \varepsilon_1^2 \varepsilon_2^3 b_1 - \hat{J}_3(\varepsilon_1^2 + \bar{\varepsilon}_1^1)\varepsilon_2^3 b_{2.1} - \hat{J}_3(\varepsilon_2^2 + \bar{\varepsilon}_1^1)\varepsilon_2^3 b_2$$
$$- [(\varepsilon_1^2 + \bar{\varepsilon}_1^1)(\hat{J}_2 \varepsilon_2^{2.1} + \hat{J}_3 \varepsilon_2^3) + \bar{\varepsilon}_2^2 \bar{\varepsilon}_3^3] b_3/|D| \tag{16.65}$$

The expressions for the concentration control coefficients for the branched pathway under consideration are listed in the appendix of chapter 15. These expressions can be substituted into the foregoing equations for the logarithmic concentrations of the intermediate metabolites.

$$y_1 = \varepsilon_0^1 C_1^{M_1} y_0 + \frac{(\hat{J}_3 C_2^{M_1} + C_3^{M_1} + C_4^{M_1})}{\hat{J}_3} b_1 + \frac{C_{2.1}^{M_1}}{\hat{J}_2} b_{2.1}$$

$$+ \frac{(C_3^{M_1} + C_4^{M_1})}{\hat{J}_3} b_2 + \frac{C_4^{M_1}}{\hat{J}_3} b_3 \tag{16.66}$$

$$y_2 = \varepsilon_0^1 C_1^{M_2} y_0 + \frac{(\hat{J}_3 C_2^{M_2} + C_3^{M_2} + C_4^{M_2})}{\hat{J}_3} b_1 + \frac{C_{2.1}^{M_2}}{\hat{J}_2} b_{2.1}$$

$$+ \frac{(C_3^{M_2} + C_4^{M_2})}{\hat{J}_3} b_2 + \frac{C_4^{M_2}}{\hat{J}_3} b_3 \tag{16.67}$$

$$y_3 = \hat{J}_3 \varepsilon_0^1 C_1^{M_3} y_0 + (\hat{J}_3 C_2^{M_3} + C_3^{M_3} + C_4^{M_3}) b_1 + \frac{\hat{J}_3 C_{2.1}^{M_3}}{\hat{J}_2} b_{2.1}$$

$$+ (C_3^{M_3} + C_4^{M_3}) b_2 + C_4^{M_3} b_3 \tag{16.68}$$

As in the case of linear multi-enzyme pathways, the equations for the logarithmic concentration of the intermediate metabolites can be written in terms of the concentration control coefficients if simple rules are adhered to with respect to the relative fluxes. These rules are the following:

1. If the metabolite for which the expression is being written is at the branch point or lies 'upstream' from the branch point, the concentration control coefficients for all of the enzymes 'downstream' from the branch point are divided by the relative flux for the branch.

2. If the metabolite for which the expression is being written lies 'downstream' from the branch point, the concentration control coefficients of all metabolites located on any other than that of the metabolite whose logarithmic concentration is being expressed should be multiplied by the relative flux appropriate for the metabolite in question.

These rules are illustrated in eqs. (16.66)–(16.68).

16.4 The future of metabolic control in biology

The present chapter together with the two previous chapters serves to introduce quantitative analysis of control of multi-enzyme systems. Although this facet of enzyme kinetics emerged two decades ago, it is still in the developmental stage. Nevertheless, it is possible to provide only a brief overview in this text. Several recent books provide more detailed discussions[4,13-15]. In the opinion of this author, this aspect of biochemistry is one of the most challenging and it holds the promise of great reward, for it provides a direct entrance into the quintessence of living systems. The enormous amount of information which has been accumulated in the past half century concerning intermediary metabolism, the refinements and extensions which have been defined in applied mathematics, and the development of molecular engineering all combine to render this approach to biology feasible at this time. Within the area of analysis of multi-enzyme systems, efforts are being directed to extend this approach from *in vitro* systems to organelles and tissues[16]. Although the temporal analysis of multi-enzyme systems[17-19] is not discussed in this textbook, treatments for the integration of temporal analysis of multi-enzyme systems and metabolic control theory have been published[20-23].

Even such complex properties of the intact animals as growth[24,25] and the response to nutrients[26,27] are becoming amenable to this type of analysis. The principles which have been developed in the analysis of multi-enzyme systems provide a foundation for initiating probes into the biochemical horizon.

16.5 Problems for chapter 16

16.1 Derive the expressions for the steady state logarithmic concentrations of the intermediate metabolites in terms of elasticity coefficients for the pathway portrayed in problem 14.1.

16.2 Express the equations derived in problem 16.1 in terms of concentration control coefficients.

16.3 Derive the expressions for the steady state logarithmic concentrations of the intermediate metabolites in terms of elasticity coefficients for the pathway shown in problem 14.4.

Appendix

There are occasions when it is convenient to express the inverse of a matrix in an element by element manner. Consider the following square matrix:

$$A = \begin{vmatrix} a_{11} & a_{12} & a_{13} \\ a_{21} & a_{22} & a_{23} \\ a_{31} & a_{32} & a_{33} \end{vmatrix}$$

The symbolic determinant of A is

$$|D| = \begin{matrix} a_{11}a_{22}a_{33} - a_{11}a_{23}a_{32} - a_{12}a_{21}a_{33} \\ + a_{12}a_{23}a_{31} + a_{13}a_{21}a_{32} - a_{13}a_{22}a_{31} \end{matrix}$$

This symbolic determinant could be rearranged as

$$\begin{matrix} a_{11}(a_{22}a_{23} - a_{23}a_{32}) & -a_{12}(a_{21}a_{23} - a_{23}a_{31}) & a_{13}(a_{21}a_{32} - a_{22}a_{31}) \\ -a_{21}(a_{12}a_{33} - a_{13}a_{32}) = & -a_{22}(a_{11}a_{33} - a_{13}a_{31}) = & -a_{23}(a_{11}a_{32} - a_{12}a_{31}) \\ a_{31}(a_{12}a_{23} - a_{13}a_{22}) & -a_{32}(a_{11}a_{23} - a_{13}a_{21}) & a_{33}(a_{11}a_{22} - a_{12}a_{21}) \end{matrix}$$

The foregoing could be expressed more concisely as

$$|D| = \begin{matrix} a_{11}M_{11} & -a_{12}M_{21} & a_{13}M_{31} \\ -a_{21}M_{12} = & a_{22}M_{22} = & -a_{23}M_{32} \\ a_{13}M_{13} & -a_{23}M_{23} & a_{33}M_{33} \end{matrix}$$

These equivalent expressions for the symbolic determinant can be expressed in the following matrices:

$$\begin{vmatrix} a_{11} & -a_{12} & a_{13} \\ a_{21} & a_{22} & -a_{23} \\ a_{31} & -a_{32} & a_{33} \end{vmatrix} \begin{vmatrix} M_{11} & M_{12} & M_{13} \\ M_{21} & M_{22} & M_{23} \\ M_{31} & M_{32} & M_{33} \end{vmatrix} = \begin{vmatrix} |D| & 0 & 0 \\ 0 & |D| & 0 \\ 0 & 0 & |D| \end{vmatrix} = |D|I$$

The M matrix in the foregoing expression is called the adjoint of matrix A, that is, M = (adj. A)[28]. The elements in the transpose of the adjoint of matrix A are the cofactors of matrix A.

$$\begin{vmatrix} \begin{vmatrix} a_{22} & a_{23} \\ a_{32} & a_{33} \end{vmatrix} & -\begin{vmatrix} a_{21} & a_{23} \\ a_{31} & a_{33} \end{vmatrix} & \begin{vmatrix} a_{21} & a_{22} \\ a_{31} & a_{32} \end{vmatrix} \\ -\begin{vmatrix} a_{12} & a_{13} \\ a_{32} & a_{33} \end{vmatrix} & \begin{vmatrix} a_{11} & a_{13} \\ a_{31} & a_{33} \end{vmatrix} & \begin{vmatrix} a_{11} & a_{12} \\ a_{31} & a_{32} \end{vmatrix} \\ \begin{vmatrix} a_{22} & a_{23} \\ a_{32} & a_{33} \end{vmatrix} & -\begin{vmatrix} a_{11} & a_{13} \\ a_{31} & a_{23} \end{vmatrix} & \begin{vmatrix} a_{11} & a_{12} \\ a_{21} & a_{22} \end{vmatrix} \end{vmatrix} = M^t = \begin{vmatrix} M_{11} & M_{21} & M_{31} \\ M_{12} & M_{22} & M_{32} \\ M_{13} & M_{23} & M_{33} \end{vmatrix}$$

It has been pointed out previously that $AA^{-1} = I$. It is also true that $A(\text{adj. A}) = |D|I$. Therefore it follows $A((\text{adj. A})/|D|)$ and $A^{-1} = ((\text{adj. A})/|D|) = M/|D|$. This provides an additional, but not a particularly convenient, means of matrix inversion. However, it shows that each element in M divided by the symbolic determinant of A is an element in the inverse matrix, and this is very useful if one wishes to perform further matrix algebra on the inverse matrix. This procedure was employed earlier in this chapter.

References

1. Savageau, M. A. (1969). Biochemical systems analysis. I Some mathematical properties of the rate law for the component enzymatic reactions. *J. Theor. Biol.* **25**: 365–69.

2. Savageau, M. A. (1972). The behavior of intact biochemical control systems. *Curr. Top. Cell. Regul.* **6**: 63–130.
3. Savageau, M. A. (1976). *Biochemical Systems Analysis: A Study of Function and Design in Molecular Biology*, Reading, Mass., Addison-Wesley Publ. Co.
4. Voit, E. O. (1991). *Canonical Nonlinear Modeling*, New York, Van Nostrand Reinhold.
5. Savageau, M. A., Voit, E. O. and Irvine, D. H. (1987). Biochemical systems theory and metabolic control theory: 1. Fundamental similarities and differences. *Math. Biosci.* **86**: 127–45.
6. Savageau, M. A., Voit, E. O. and Irvine, D. H. (1987). Biochemical systems theory and metabolic control theory: 2. The role of summation and connectivity relationships. *Math. Biosci.* **86**: 147–69.
7. Sen, A. K. and Schulz, A. R. (1989). Effects of product inhibition in metabolic pathways: Stability and control. *Math. Biosci.* **96**: 255–77.
8. Voit, E. O. and Savageau, M. A. (1987). Accuracy of alternative representations for integrated biochemical systems. *Biochemistry* **26**: 6869–80.
9. Atkinson, D. E. (1969). Limitation of metabolite concentration and conservation of solvent capacity in the living cell. *Curr. Top. Cell. Regul.* **1**: 29–43.
10. Srivastava, D. K. and Bernhard, S. A. (1987). Biophysical chemistry of metabolic reaction sequences in concentrated enzyme solution and in the cell. *Ann. Rev. Biophys. Chem.* **16**: 175–204.
11. Mendes, P., Kell, D. B. and Westerhoff, H. V. (1992). Channelling can decrease pool size. *Eur. J. Biochem.* **204**: 257–66.
12. Kacser, H., Sauro, H. M. and Acerenza, L. (1990). Enzyme-enzyme interactions and control analysis. *Eur. J. Biochem.* **187**: 481–91.
13. Cornish-Bowden, A. and Cardenas, M. L. (1989). *Control of Metabolic Processes*, New York, Plenum Press.
14. Westerhoff, H. V. and van Dam, K. (1987). *Thermodynamics and Control of Biological Free-energy Transduction*, Amsterdam, Elsevier.
15. Hill, T. L. (1989) *Free Energy Transduction and Biochemical Cycle Kinetics*, New York, Springer-Verlag.
16. Brown, G. C., Hafner, R. P. and Brand, M. D. (1990). A 'top-down' approach to the determination of control coefficients in metabolic control theory. *Eur. J. Biochem.* **188**: 321–25.
17. Easterby, J. S. (1973). Coupled enzyme assays: A general expression for the transient. *Biochim. Biophys. Acta* **293**: 552–58.
18. Easterby, J. S. (1981). Generalized theory of the transition time for sequential enzyme reactions. *Biochem. J.* **199**: 155–61.
19. Easterby, J. S. (1984). The kinetics of consecutive enzyme reactions. *Biochem. J.* **219**: 843–47.
20. Ovadi, J., Tompa, P., Vertessy, B., Orosz, F., Keleti, T. and Welch, G. R. (1989). Transient-time analysis of substrate-channelling in interacting enzyme systems. *Biochem. J.* **257**: 187–90.
21. Torres, N. V., Souto, R. and Melendez-Hevia, E. (1989). Study of the flux and transition time control coefficient profiles in a metabolic system in vitro and the effect of an external stimulator. *Biochem. J.* **260**: 763–69.
22. Easterby, J. S. (1990). Integration of temporal analysis and control analysis of metabolic systems. *Biochem. J.* **269**: 255–59.
23. Torres, N. V., Sicilia, J. and Melendez-Hevia, E. (1991). Analysis and characterization of transition states in metabolic systems. *Biochem. J.* **276**: 231–36.

24. Savageau, M. A. (1979). Growth of complex systems can be related to the properties of their underlying determinants. *Proc. Natl. Acad. Sci. USA* **76**: 5413–417.
25. Savageau, M. A. (1980). Growth equations: A general equation and a survey of special cases. *Math. Biosci.* **48**: 267–78.
26. Schulz, A. R. (1991). Interpretation of nutrient-response relationships in rats. *J. Nutr.* **121**: 1834–43.
27. Schulz, A. R. (1992). Nutrient-response: A long random walk through metabolic pathways. *J. Nutr. Biochem.* **3**: 98–103.
28. Magar, M. E. (1972). *Data Analysis in Biochemistry and Biophysics.*, pp. 20–54, New York, Academic Press.

Part Five

Solutions to problems

Chapter 1

1.1 $\dfrac{d(EA)}{dt} = k_1(E)(A) - (k_{-1} + k_2)(EA) = 0$ $E_t = (E) + (EA)$

$(E) = \dfrac{(k_{-1} + k_2)(EA)}{k_1(A)} = \dfrac{K_m(EA)}{(A)}$

$(EA) = \dfrac{k_1(A)(E)}{k_{-1} + k_2} = \dfrac{(A)(E)}{K_m}$

$\dfrac{(E)}{E_t} = \left[1 + \dfrac{(A)}{K_m} \right]^{-1}$ $\dfrac{(EA)}{E_t} = \left[1 + \dfrac{K_m}{(A)} \right]^{-1}$

1.2

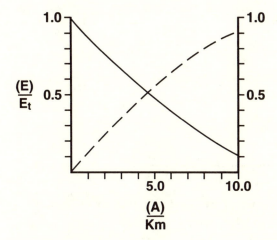

Fig. V.1.1. Plots of $(E)/E_t$ and $(EA)/E_t$ versus $(A)/K_m$ for a simple enzymic reaction.

1.3 $K_m = 0.323\,mM$ $V_{max} = 0.0612\,\mu moles/minute$

Chapter 2

2.1 Saturate the enzyme with substrate A.

$k_2 = \dfrac{V_{max}}{E_t}$

Follow the time-course of the enzyme-catalyzed reaction at a sub-saturating concentration of A from the pre-steady state into the steady phase of the reaction.

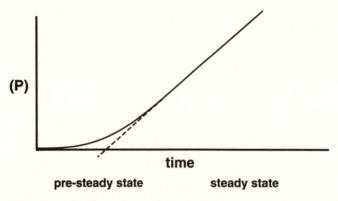

(P)

time

pre-steady state   **steady state**

Fig. V.2.1. Time-course curve for a simple enzymic reaction. The time-course curve is shown through the pre-steady state and steady state phases. The steady state segment is extrapolated to the time axis.

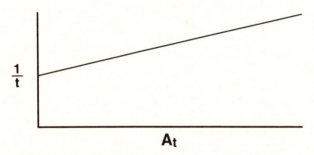

$\frac{1}{t}$

A$_t$

Fig. V.2.2. Plot of the reciprocal of the intersection of the time axis of Figure V.2.1 versus A$_t$.

The coordinate of the point at which the extrapolated line intersects the time axis is,

$$t = \frac{1}{k_{-1} + k_2 + k_1 A_t}$$

Repeat the time-course study at various concentrations of A and plot $1/t$ versus A$_t$.

The slope of the plot is equal to k_1, and the point of intersection of the $1/t$ axis is $k_{-1} + k_2$. Since k_2 is known, k_{-1} can be calculated.

2.2 $75,000 \, X = \dfrac{100,000 \times 0.05 \, K_m}{K_m + 0.05 \, K_m}$

$X = 6.3\%$.

2.3 $25{,}000\,X = \dfrac{100{,}000 \times 0.05\,K_m}{K_m + 0.05\,K_m}$

$X = 19\%.$

Chapter 3

3.1 $(I) = 0.000\,mM \quad K_m^{app} = 0.204\,mM \quad V_{max}^{app} = 13.50\,\mu moles/min.$
$(I) = 1.393\,mM \quad K_m^{app} = 0.202\,mM \quad V_{max}^{app} = 9.144\,\mu moles/min.$
$(I) = 2.790\,mM \quad K_m^{app} = 0.207\,mM \quad V_{max}^{app} = 7.706\,\mu moles/min.$
$(I) = 4.180\,mM \quad K_m^{app} = 0.197\,mM \quad V_{max}^{app} = 6.221\,\mu moles/min.$
$(I) = 5.570\,mM \quad K_m^{app} = 0.208\,mM \quad V_{max}^{app} = 5.844\,\mu moles/min.$

true $K_m = 0.210\,mM$ true $V_{max} = 12.51\,\mu moles/min.$
slope inhibition constant $= K_{is} = 4.650\,mM$
intercept inhibition constant $= K_{ii} = 4.530\,mM$

3.2 $\dfrac{v_0}{v_i} = \dfrac{K_m\left[1 + \dfrac{(I)}{\bar{\bar{K}}_3}\right] + (A)\left[1 + \dfrac{(I)}{\bar{K}_4}\right]}{K_m + (A)}$

$\dfrac{v_0 - v_i}{v_i} = \dfrac{\left[\dfrac{K_m}{\bar{K}_3} + \dfrac{(A)}{\bar{K}_4}\right]}{k_m + (A)}(I)$

$(I) = \dfrac{\bar{K}_3\bar{K}_4\left[1 + \dfrac{K_m}{(A)}\right]}{\bar{K}_3 + \bar{K}_4\dfrac{K_m}{(A)}}\left(\dfrac{v_0 - v_i}{v_i}\right)$

3.3a Note that when $v_i = 0.5 \times v_0,\ \left(\dfrac{v_0 - v_i}{v_i}\right) = 1.$

$(I) = \bar{K}_i\left(\dfrac{v_0 - v_i}{v_i}\right)$

3.3b $(I) = \bar{K}_4\left(\dfrac{v_0 - v_1}{v_i}\right)$

3.3c $(I) = \bar{K}_3\left[1 + \dfrac{(A)}{K_m}\right]\left(\dfrac{v_0 - v_i}{v_i}\right)$

3.3d $(I) = 2\bar{K}_3 \left(\dfrac{v_0 - v_i}{v_i} \right)$

3.3e $(I) = \bar{K}_4 \left[1 + \dfrac{K_m}{(A)} \right] \left(\dfrac{v_0 - v_i}{v_i} \right)$

3.3f $(I) = \bar{K}_4 \left(\dfrac{v_0 - v_i}{v_1} \right)$

Chapter 4

4.1 $E_t = (E) + (EA)$

$$\frac{d(E)}{dt} = -[k_1(A) + k_{-2}(P)](E) + (k_{-1} + k_2)(EA) = 0$$

$$\frac{d(EA)}{dt} = [k_1(A) + k_{-2}(P)](E) + (k_{-1} + k_2)(EA) = 0$$

$$v = \left[k_2 \frac{(EA)}{E_t} - k_{-2} \frac{(E)}{E_t}(P) \right] E_t$$

$$\frac{(EA)}{E_t} = \frac{k_1(A) + k_{-2}(P)}{k_{-1} + k_2 + k_1(A) + k_{-2}(P)}$$

$$\frac{(E)}{E_t} = \frac{k_{-1} + k_2}{k_{-1} + k_2 + k_1(A) + k_{-2}(P)}$$

$$v = \frac{[k_1 k_2(A) - k_{-1} k_{-2}(P)] E_t}{k_{-1} + k_2 + k_1(A) + k_{-2}(P)}$$

$$V_f = k_2 E_t \qquad V_r = k_{-1} E_t$$

$$K_a = \frac{k_{-1} + k_2}{k_1} \quad K_p = \frac{k_{-1} + k_2}{k_{-2}} \quad K_{eq} = \frac{k_1 k_2}{k_{-1} k_{-2}}$$

$$v = \frac{V_f \left[(A) - \dfrac{(P)}{K_{eg}} \right] E_t}{K_a + (A) + \dfrac{K_a}{K_p}(P)}$$

4.2

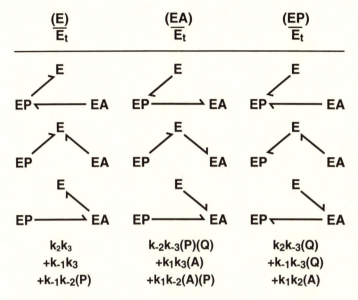

Fig. V.4.1 King-Altman spanning trees for the enzymic reaction sequence in problem 4.2.

Chapter 5

5.1a The connection matrix for the mechanism is,

$$U = \begin{vmatrix} 0 & A & 0 & 0 & Q \\ 1 & 0 & 1 & 0 & 0 \\ 0 & P & 0 & B & 0 \\ 0 & 0 & 1 & 0 & C \\ 1 & 0 & 0 & 1 & 0 \end{vmatrix}$$

$$Q_e = \begin{vmatrix} 0 & 0 \\ 1 & 3 \\ 2 & 4 \\ 3 & 5 \\ 1 & 4 \end{vmatrix} \doteq \begin{matrix} (0,1,2,3,1) & (P) \\ (0,1,2,3,4) & (P) \\ (0,1,2,5,1) & (C) & (P) \\ (0,1,4,5,1) & (B) & (C) \\ (0,3,4,5,1) & (B) & (C) \end{matrix} = \frac{(E)}{E_t}$$

$$Q_{ea} = \begin{vmatrix} 2 & 5 \\ 0 & 0 \\ 2 & 4 \\ 3 & 5 \\ 1 & 4 \end{vmatrix} \doteq \begin{array}{l} (2,0,2,3,1)\ (A)\ (P) \\ (2,0,2,3,4)\ (A)\ (P) \\ (2,0,2,5,1)\ (A)\ (C)\ (P) \\ (2,0,4,5,1)\ (A)\ (B)\ (C) \\ (5,0,2,3,4)\ (P)\ (Q) \end{array} = \frac{(EA)}{E_t}$$

$$Q_{f} = \begin{vmatrix} 2 & 5 \\ 1 & 3 \\ 0 & 0 \\ 3 & 5 \\ 1 & 4 \end{vmatrix} \doteq \begin{array}{l} (2,3,0,3,1)\ (A) \\ (2,3,0,3,4)\ (A) \\ (2,3,0,5,1)\ (A)\ (C) \\ (5,1,0,3,4)\ (Q) \\ (5,3,0,3,4)\ (Q) \end{array} = \frac{(F)}{E_t}$$

$$Q_{eb} = \begin{vmatrix} 2 & 5 \\ 1 & 3 \\ 2 & 4 \\ 0 & 0 \\ 1 & 4 \end{vmatrix} \doteq \begin{array}{l} (2,3,4,0,1)\ (A)\ (B) \\ (2,3,4,0,4)\ (A)\ (B) \\ (5,1,2,0,4)\ (P)\ (Q) \\ (5,1,4,0,4)\ (B)\ (Q) \\ (5,3,4,0,4)\ (B)\ (Q) \end{array} = \frac{(EB)}{E_t}$$

$$Q_{eq} = \begin{vmatrix} 2 & 5 \\ 1 & 3 \\ 2 & 4 \\ 3 & 5 \\ 0 & 0 \end{vmatrix} \doteq \begin{array}{l} (2,3,4,5,0)\ (A)\ (B)\ (C) \\ (5,1,2,3,0)\ (P)\ (Q) \\ (5,1,2,5,0)\ (C)\ (P)\ (Q) \\ (5,1,4,5,0)\ (B)\ (C)\ (Q) \\ (5,3,4,5,0)\ (B)\ (C)\ (Q) \end{array} = \frac{(EQ)}{E_t}$$

$$v = \frac{[(2,3,4,5,1)(A)(B)(C) - (5,1,2,3,4)(P)(Q)]E_t}{\begin{array}{llll} (2,3,0,3,1)(A) & (2,3,4,0,1)(A)(B) & (5,1,4,0,4)(B)(Q) & (2,3,4,5,0)(A)(B)(C) \\ (2,3,0,3,4)(A) & (2,3,4,0,4)(A)(B) & (5,3,4,0,4)(B)(Q) & (2,0,4,5,1)(A)(B)(C) \\ (0,1,2,3,1)(P) & (2,3,0,5,1)(A)(C) & (0,1,2,5,1)(C)(P) & (2,0,4,5,1)(A)(C)(P) \\ (0,1,2,3,4)(P) & (2,0,2,3,1)(A)(P) & (5,1,2,3,0)(P)(Q) & (5,1,4,5,0)(B)(C)(Q) \\ (5,1,0,3,4)(Q) & (2,0,2,3,4)(A)(P) & (5,1,2,0,4)(P)(Q) & (5,3,4,5,0)(B)(C)(Q) \\ (5,3,0,3,4)(Q) & (0,1,4,5,1)(B)(C) & (5,0,2,3,4)(P)(Q) & (5,1,2,5,0)(C)(P)(Q) \\ & (0,3,4,5,1)(B)(C) & & \end{array}}$$

5.1b $\quad V_f = \dfrac{\text{num. } 1}{\text{coef. ABC}} = \dfrac{(2,3,4,5,1)\,E_t}{\begin{array}{c}(2,3,4,5,0) \\ (2,0,4,5,1)\end{array}} = \dfrac{k_{23}k_{51}E_t}{k_{23}+k_{51}}$

$$V_r = \frac{\text{num. 2}}{\text{coef. PQ}} = \frac{\begin{array}{c}(5,1,2,3,4)\,E_t\end{array}}{\begin{array}{c}(5,1,2,3,0)\\(5,1,2,0,4)\\(5,0,2,3,4)\end{array}} = \frac{k_{21}k_{43}k_{54}E_t}{k_{21}k_{43}+k_{21}k_{54}+k_{43}k_{54}}$$

$$K_a = \frac{\text{coef. BC}}{\text{coef. ABC}} = \frac{\begin{array}{c}(0,1,4,5,1)\\(0,3,4,5,1)\end{array}}{\begin{array}{c}(2,3,4,5,0)\\(2,0,4,5,1)\end{array}} = \frac{k_{51}(k_{21}+k_{23})}{k_{12}(k_{23}+k_{51})}$$

$$K_b = \frac{\text{coef. AC}}{\text{coef. ABC}} = \frac{(2,3,0,5,1)}{\begin{array}{c}(2,3,4,5,0)\\(2,0,4,5,1)\end{array}} = \frac{k_{23}k_{51}}{k_{34}(k_{23}+k_{51})}$$

$$K_c = \frac{\text{coef. AB}}{\text{coef. ABC}} = \frac{\begin{array}{c}(2,3,4,0,1)\\(2,3,4,0,4)\end{array}}{\begin{array}{c}(2,3,4,5,0)\\(2,0,4,5,1)\end{array}} = \frac{k_{23}(k_{51}+k_{54})}{k_{45}(k_{23}+k_{51})}$$

$$K_p = \frac{\text{coef. Q}}{\text{coef. PQ}} = \frac{\begin{array}{c}(5,1,0,3,4)\\(5,3,0,3,4)\end{array}}{\begin{array}{c}(5,1,2,3,0)\\(5,1,2,0,4)\\(5,0,2,3,4)\end{array}} = \frac{k_{43}k_{54}(k_{21}+k_{23})}{k_{32}(k_{21}k_{43}+k_{21}k_{54}+k_{43}k_{54})}$$

$$K_q = \frac{\text{coef. P}}{\text{coef. PQ}} = \frac{\begin{array}{c}(0,1,2,3,1)\\(0,1,2,3,4)\end{array}}{\begin{array}{c}(5,1,2,3,0)\\(5,1,2,0,4)\\(5,0,2,3,4)\end{array}} = \frac{k_{21}k_{43}(k_{51}+k_{54})}{k_{15}(k_{21}k_{43}+k_{21}k_{54}+k_{43}k_{54})}$$

$$K_{ia} = \frac{\text{coef. P}}{\text{coef. AP}} = \frac{\begin{array}{c}(0,1,2,3,1)\\(0,1,2,3,4)\end{array}}{\begin{array}{c}(2,0,2,3,1)\\(2,0,2,3,4)\end{array}} = \frac{\text{coef. CP}}{\text{coef. ACP}} = \frac{(0,1,2,5,1)}{(2,0,2,5,1)} = \frac{k_{21}}{k_{12}}$$

$$K_{ib} = \frac{\text{coef. A}}{\text{coef. AB}} = \frac{\begin{array}{c}(2,3,0,3,1)\\(2,3,0,3,4)\end{array}}{\begin{array}{c}(2,3,4,0,1)\\(2,3,4,0,4)\end{array}} = \frac{\text{coef. Q}}{\text{coef. BQ}} = \frac{\begin{array}{c}(5,1,0,3,4)\\(5,3,0,3,4)\end{array}}{\begin{array}{c}(5,1,4,0,4)\\(5,3,4,0,4)\end{array}} = \frac{k_{43}}{k_{34}}$$

$$K_{ic} = \frac{\text{coef. BQ}}{\text{coef. BCQ}} = \frac{\begin{matrix}(5,1,4,0,4)\\(5,3,4,0,4)\\(5,1,4,5,0)\\(5,3,4,5,0)\end{matrix}}{} = \frac{k_{54}}{k_{45}}$$

$$K_{ip} = \frac{\text{coef. A}}{\text{coef. AP}} = \frac{\begin{matrix}(2,3,0,3,1)\\(2,3,0,3,4)\\(2,0,2,3,1)\\(2,0,2,3,4)\end{matrix}}{} = \frac{\text{coef. AC}}{\text{coef. ACP}} = \frac{(2,3,0,5,1)}{(2,0,2,5,1)} = \frac{k_{23}}{k_{32}}$$

$$K_{iq} = \frac{\text{coef. BC}}{\text{coef. BCQ}} = \frac{\begin{matrix}(0,1,4,5,1)\\(0,3,4,5,1)\\(5,1,4,5,0)\\(5,3,4,5,0)\end{matrix}}{} = \frac{\text{coef. CP}}{\text{coef. CPQ}} = \frac{(0,1,2,5,1)}{(5,1,2,5,0)} = \frac{k_{51}}{k_{15}}$$

$$K_{eq} = \frac{\text{num. 1}}{\text{num. 2}} = \frac{(2,3,4,5,1)E_t}{(5,1,2,3,4)E_t} = \frac{k_{12}k_{23}k_{34}k_{45}k_{51}}{k_{15}k_{21}k_{32}k_{43}k_{54}}$$

5.1c

$$v = \frac{V_f\left[(A)(B)(C) - \dfrac{(P)(Q)}{K_{eq}}\right]}{\begin{aligned}&K_{ib}K_c(A) + K_c(A)(B) + K_b(A)(C) + K_a(B)(C)\\[4pt]&+ (A)(B)(C) + \frac{K_{ib}K_c}{K_{ip}}(A)(P) + \frac{K_aK_{ic}}{K_{iq}}(B)(Q)\\[4pt]&+ \frac{K_{ia}K_b}{K_{ip}}(C)(P) + \frac{K_b}{K_{ip}}(A)(C)(P) + \frac{K_a}{K_{iq}}(B)(C)(Q)\\[4pt]&+ \frac{K_{ia}K_{ib}K_c}{K_{ip}}(P) + \frac{K_aK_{ib}K_{ic}}{K_{iq}}(Q) + \frac{K_aK_{ib}K_{ic}}{K_pK_{iq}}(P)(Q)\\[4pt]&+ \frac{K_{ia}K_b}{K_{ip}K_{iq}}(C)(P)(Q)\end{aligned}}$$

5.2a   The connection matrix for the mechanism is,

$$U = \begin{vmatrix} 0 & A & 0 & 0 & R \\ 1 & 0 & B & 0 & 0 \\ 0 & 1 & 0 & 1 & 0 \\ 0 & 0 & P & 0 & 1 \\ 1 & 0 & 0 & Q & 0 \end{vmatrix} \qquad Q = \begin{vmatrix} 2 & 5 \\ 1 & 3 \\ 2 & 4 \\ 3 & 5 \\ 1 & 4 \end{vmatrix}$$

$$\frac{(E)}{E_t} = \begin{array}{l} (0,1,2,3,1)\ (P) \\ (0,1,2,3,4)\ (P)\ (Q) \\ (0,1,2,5,1) \\ (0,1,4,5,1) \\ (0,3,4,5,1)\ (B) \end{array} \qquad \frac{(EA)}{E_t} = \begin{array}{l} (2,0,2,3,1)\ (A)\ (P) \\ (2,0,2,3,4)\ (A)\ (P)\ (Q) \\ (2,0,2,5,1)\ (A) \\ (2,0,4,5,1)\ (A) \\ (5,0,2,3,4)\ (P)\ (Q)\ (R) \end{array}$$

$$\frac{(EAB)}{E_t} = \begin{array}{l} (2,3,0,3,1)\ (A)\ (B)\ (P) \\ (2,3,0,3,4)\ (A)\ (B)\ (P)\ (Q) \\ (2,3,0,5,1)\ (A)\ (B) \\ (5,1,0,3,4)\ (P)\ (Q)\ (R) \\ (5,3,0,3,4)\ (B)\ (P)\ (Q)\ (R) \end{array}$$

$$\frac{(EQR)}{E_t} = \begin{array}{l} (2,3,4,0,1)\ (A)\ (B) \\ (2,3,4,0,4)\ (A)\ (B)\ (Q) \\ (5,1,2,0,4)\ (Q)\ (R) \\ (5,1,4,0,4)\ (Q)\ (R) \\ (5,3,4,0,4)\ (B)\ (Q)\ (R) \end{array} \qquad \frac{(ER)}{E_t} = \begin{array}{l} (2,3,4,5,0)\ (A)\ (B) \\ (5,1,2,3,0)\ (P)\ (R) \\ (5,1,2,5,0)\ (R) \\ (5,1,2,4,0)\ (R) \\ (5,3,4,5,0)\ (B)\ (R) \end{array}$$

$$v = \frac{[(2,3,4,5,1)(A)(B) - (5,1,2,3,4)(P)(Q)(R)]E_t}{\begin{array}{lll} (0,1,2,5,1) & (2,3,4,5,0)(A)(B) & (2,3,0,3,1)(A)(B)(P) \\ (0,1,4,5,1) & (2,3,4,0,1)(A)(B) & (2,3,4,0,4)(A)(B)(Q) \\ & (2,3,0,5,1)(A)(B) & \\ (2,0,2,5,1)(A) & & (2,0,2,3,4)(A)(P)(Q) \\ (2,0,4,5,1)(A) & (2,0,2,3,1)(A)(P) & (5,3,4,0,4)(B)(Q)(R) \\ (0,3,4,5,1)(B) & (5,3,4,5,0)(B)(R) & (5,1,0,3,4)(P)(Q)(R) \\ (0,1,2,3,1)(P) & (0,1,2,3,4)(P)(Q) & (5,0,2,3,4)(P)(Q)(R) \\ (5,1,2,5,0)(R) & (5,1,2,3,0)(P)(R) & (2,3,0,3,4)(A)(B)(P)(Q) \\ (5,1,4,5,0)(R) & (5,1,2,0,4)(Q)(R) & (5,3,0,3,4)(B)(P)(Q)(R) \\ & (5,1,4,0,4)(Q)(R) & \end{array}}$$

5.2b $$V_f = \frac{(2,3,4,5,1)\,E_t}{\begin{array}{l} (2,3,4,5,0) \\ (2,3,4,0,1) \\ (2,3,0,4,1) \end{array}} = \frac{k_{34}k_{45}k_{51}\,E_t}{k_{34}k_{45} + k_{34}k_{51} + k_{45}k_{51}}$$

$$V_r = \frac{(5,1,2,3,4)\,E_t}{\begin{array}{l} (5,1,0,3,4) \\ (5,0,2,3,4) \end{array}} = \frac{k_{21}k_{32}\,E_t}{k_{21} + k_{32}}$$

$$K_a = \frac{\begin{array}{l}(0,3,4,5,1)\\(2,3,4,5,0)\\(2,3,4,0,1)\\(2,3,0,5,1)\end{array}}{} = \frac{k_{34}k_{45}k_{51}}{k_{12}(k_{34}k_{45} + k_{34}k_{51} + k_{45}k_{51})}$$

$$K_b = \frac{\begin{array}{l}(2,0,2,5,1)\\(2,0,4,5,1)\\(2,3,4,5,0)\\(2,3,4,0,1)\\(2,3,0,5,1)\end{array}}{} = \frac{k_{45}k_{51}(k_{32} + k_{34})}{k_{23}(k_{34}k_{45} + k_{34}k_{51} + k_{45}k_{51})}$$

$$K_p = \frac{\begin{array}{l}(5,1,2,0,4)\\(5,1,4,0,4)\\(5,1,0,3,4)\\(5,0,2,3,4)\end{array}}{} = \frac{k_{21}(k_{32} + k_{34})}{k_{43}(k_{21} + k_{32})}$$

$$K_q = \frac{\begin{array}{l}(5,1,2,3,0)\\(5,1,0,3,4)\\(5,0,2,3,4)\end{array}}{} = \frac{k_{21} + k_{32}}{k_{54}(k_{21} + k_{32})}$$

$$K_r = \frac{\begin{array}{l}(0,1,2,3,4)\\(5,1,0,3,4)\\(5,0,2,3,4)\end{array}}{} = \frac{k_{21}k_{32}}{k_{15}(k_{21} + k_{32})}$$

$$K_{ia} = \frac{\begin{array}{l}(0,1,2,5,1)\\(0,1,4,5,1)\\(2,0,2,5,1)\\(2,0,4,5,1)\end{array}}{} = \frac{(0,1,2,3,1)}{(2,0,2,3,1)} = \frac{(0,1,2,3,4)}{(2,0,2,3,4)} = \frac{k_{21}}{k_{12}}$$

$$K_{ip_1} = \frac{\begin{array}{l}(2,0,2,5,1)\\(2,0,4,5,1)\\(2,0,2,3,1)\end{array}}{} = \frac{\begin{array}{l}(5,1,2,5,0)\\(5,1,4,5,0)\\(5,1,2,3,0)\end{array}}{} = \frac{k_{45}(k_{32} + k_{34})}{k_{32}k_{43}}$$

$$K_{ip_2} = \frac{\begin{array}{l}(2,3,4,5,0)\\(2,3,4,0,1)\\(2,3,0,5,1)\\(2,3,0,3,1)\end{array}}{} = \frac{(k_{34}k_{45} + k_{34}k_{51} + k_{45}k_{51})}{k_{43}k_{51}}$$

$$K_{ip_3} = \frac{(5,3,4,0,4)}{(5,3,0,3,4)} = \frac{k_{34}}{k_{43}}$$

$$K_{iq_1} = \frac{(0,1,2,3,1)}{(0,1,2,3,4)} = \frac{(2,0,32,3,1)}{(2,0,2,3,4)} = \frac{(2,3,0,3,1)}{(2,3,0,3,4)} = \frac{k_{51}}{k_{54}}$$

$$K_{iq_2} = \frac{\begin{array}{c}(2,3,4,5,0)\\(2,3,4,0,1)\\(2,3,0,5,1)\end{array}}{(2,3,4,0,4)} = \frac{k_{34}k_{45} + k_{34}k_{51} + k_{45}k_{51}}{k_{34}k_{54}}$$

$$K_{iq_3} = \frac{\begin{array}{c}(5,1,2,5,0)\\(5,1,4,5,0)\\(5,1,2,0,4)\\(5,1,4,0,4)\end{array}}{\begin{array}{c}(5,3,4,5,0)\\(5,3,4,0,4)\end{array}} = \frac{k_{45}}{k_{54}}$$

$$K_{ir} = \frac{\begin{array}{c}(0,1,2,5,1)\\(0,1,4,5,1)\\(5,1,2,5,0)\\(5,1,4,5,0)\end{array}}{\begin{array}{c}(0,3,4,5,1)\\(5,3,4,5,0)\end{array}} = \frac{k_{51}}{k_{15}}$$

5.2c

$$v = \frac{V_f\left[(A)(B) - \dfrac{(P)(Q)(R)}{K_{eq}}\right]}{\begin{array}{l} K_{ia}K_b + K_b(A) + K_a(B) + \dfrac{K_{ia}K_b}{K_{ip_1}}(P) \\[2ex] + \dfrac{K_{ia}K_b}{K_{ir}}(R) + (A)(B) + \dfrac{K_b}{K_{ip_1}}(A)(P) + \dfrac{K_a}{K_{ir}}(B)(R) \\[2ex] + \dfrac{K_{ia}K_b}{K_{ip_1}K_{iq_1}}(P)(Q) + \dfrac{K_{ia}K_b}{K_{ip_1}K_{ir}}(P)(R) + \dfrac{K_{ia}K_b}{K_{iq_3}K_{ir}}(Q)(R) \\[2ex] + \dfrac{(A)(B)(P)}{K_{ip_2}} + \dfrac{(A)(B)(Q)}{K_{iq_2}} + \dfrac{K_b}{K_{ip_1}K_{iq_1}}(A)(P)(Q) \\[2ex] + \dfrac{K_a}{K_{iq_3}K_{ir}}(B)(Q)(R) + \dfrac{K_{ia}K_b}{K_p K_{iq_3}K_{ir}}(P)(Q)(R) \\[2ex] + \dfrac{(A)(B)(P)(Q)}{K_{ip_1}K_{iq_2}} + \dfrac{K_a}{K_{ip_3}K_{iq_3}K_{ir}}(B)(P)(Q)(R) \end{array}}$$

Chapter 6

6.1
$$\frac{1}{v_a} = \frac{K_a}{V_f}\left[1 + \frac{K_{ia}K_b\,(P)}{K_aK_{ip}\,(B)}\left(1 + \frac{K_{ib}K_c}{K_b(C)}\right)\right]\frac{1}{(A)}$$

$$+ \frac{1}{V_f}\left[1 + \frac{K_b}{(B)}\left(1 + \frac{(P)}{K_{ip}}\right) + \frac{K_c}{(C)}\left(1 + \frac{K_{ib}}{(B)}\left(1 + \frac{(P)}{K_{ip}}\right)\right)\right]$$

$$\frac{1}{v_b} = \frac{K_b}{V_f}\left[1 + \frac{K_{ib}K_c}{K_b(C)}\left(1 + \frac{(P)}{K_{ip}}\left(1 + \frac{K_{ia}}{(A)}\right)\right) + \frac{(P)}{K_{ip}}\left(1 + \frac{K_{ia}}{(A)}\right)\right]\frac{1}{(B)}$$

$$+ \frac{1}{V_f}\left[1 + \frac{K_a}{(A)} + \frac{K_c}{(C)}\right]$$

$$\frac{1}{v_c} = \frac{K_c}{V_f}\left[1 + \frac{K_{ib_1}}{(B)}\left(1 + \frac{(P)}{K_{ip}}\left(1 + \frac{K_{ia}}{(A)}\right)\right)\right]\frac{1}{(C)}$$

$$+ \frac{1}{V_f}\left[1 + \frac{K_a}{(A)} + \frac{K_b}{(B)}\left(1 + \frac{(P)}{K_{ip}}\left(1 + \frac{K_{ia}}{(A)}\right)\right)\right]$$

$$\frac{1}{v_a} = \frac{K_a}{V_f}\left[1 + \frac{(Q)}{K_{iq}}\left(1 + \frac{K_{ic}}{(C)}\right) + \frac{K_{ib}K_{ic}}{K_{iq}}\frac{(Q)}{(B)(C)}\right]\frac{1}{(A)}$$

$$+ \frac{1}{V_f}\left[1 + \frac{K_b}{(B)} + \frac{K_c}{(C)}\left(1 + \frac{K_{ib}}{(B)}\right)\right]$$

$$\frac{1}{v_b} = \frac{K_b}{V_f}\left[1 + \frac{K_{ib}K_c}{K_b(C)}\left(1 + \frac{K_aK_{ic}(Q)}{K_cK_{iq}(A)}\right)\right]\frac{1}{(B)}$$

$$+ \frac{1}{V_f}\left[1 + \frac{K_a}{(A)}\left(1 + \frac{(Q)}{K_{iq}}\right) + \frac{K_c}{(C)}\left(1 + \frac{K_aK_{ic}(Q)}{K_cK_{iq}(A)}\right)\right]$$

$$\frac{1}{v_c} = \frac{K_c}{V_f}\left[1 + \frac{K_{ib}}{(B)} + \frac{K_aK_{ic}}{K_cK_{iq}}\frac{(Q)}{(A)}\left(1 + \frac{K_{ib}}{(B)}\right)\right]\frac{1}{(C)}$$

$$+ \frac{1}{V_f}\left[1 + \frac{K_a}{(A)}\left(1 + \frac{(Q)}{K_{iq}}\right) + \frac{K_b}{(B)}\right]$$

6.2

(A)	(B)	(C)	Inhibitor	Type of Inhibition
Variable	Subsaturate	Subsaturate	P	Mixed type
Variable	Saturate	Subsaturate	P	No inhibition
Variable	Subsaturate	Saturate	P	Mixed type

Subsaturate	Variable	Subsaturate	P	Competitive
Saturate	Variable	Subsaturate	P	Competitive
Subsaturate	Variable	Saturate	P	Competitive
Subsaturate	Subsaturate	Variable	P	Mixed type
Saturate	Subsaturate	Variable	P	Mixed type
Subsaturate	Saturate	Variable	P	No inhibition
Variable	Subsaturate	Subsaturate	Q	Competitive
Variable	Saturate	Subsaturate	Q	Competitive
Variable	Subsaturate	Saturate	Q	Competitive
Subsaturate	Variable	Subsaturate	Q	Mixed type
Saturate	Variable	Subsaturate	Q	No inhibition
Subsaturate	Variable	Saturate	Q	Uncompetitive
Subsaturate	Subsaturate	Variable	Q	Mixed type
Saturate	Subsaturate	Variable	Q	No inhibition
Subsaturate	Saturate	Variable	Q	Mixed type

6.3

$$\frac{1}{V_a} = \frac{K_a}{V_f}\left[1 + \frac{K_{ia}K_b}{K_a(B)}\left(1 + \frac{(P)}{K_{ip_1}}\right)\right]\frac{1}{(A)}$$

$$+ \frac{1}{V_f}\left[1 + \frac{K_b}{(B)}\left(1 + \frac{(P)}{K_{ip_1}}\right) + \frac{(P)}{K_{ip_2}}\right]$$

$$\frac{1}{v_b} = \frac{K_b}{V_f}\left[1 + \frac{K_{ia}}{(A)}\left(1 + \frac{(P)}{K_{ip_1}}\right) + \frac{(P)}{K_{ip_1}}\right]\frac{1}{(B)}$$

$$+ \frac{1}{V_f}\left[1 + \frac{K_a}{(A)} + \frac{(P)}{K_{ip_2}}\right]$$

$$\frac{1}{v_a} = \frac{K_a}{V_f}\left[1 + \frac{K_{ia}K_b}{K_a(B)}\right]\frac{1}{(A)} + \frac{1}{V_f}\left[1 + \frac{K_b}{(B)} + \frac{(Q)}{K_{iq_2}}\right]$$

$$\frac{1}{v_b} = \frac{K_b}{V_f}\left[1 + \frac{K_{ia}}{(A)}\right]\frac{1}{(B)} + \frac{1}{V_f}\left[1 + \frac{K_a}{(A)} + \frac{(Q)}{K_{iq_2}}\right]$$

$$\frac{1}{v_a} = \frac{K_a}{V_f}\left[1 + \frac{K_{ia}K_b}{K_a(B)}\left(1 + \frac{(R)}{K_{ir}}\right) + \frac{(R)}{K_{ir}}\right]\frac{1}{(A)} + \frac{1}{V_f}\left[1 + \frac{K_b}{(B)}\right]$$

$$\frac{1}{v_b} = \frac{K_b}{V_f}\left[1 + \frac{K_{ia}}{(A)}\left(1 + \frac{(R)}{K_{ir}}\right)\right]\frac{1}{(B)} + \frac{1}{V_f}\left[1 + \frac{K_a}{(A)}\left(1 + \frac{(R)}{K_{ir}}\right)\right]$$

6.4

(A)	(B)	Product	Type of inhibition
Variable	Substrate	P	Mixed type
Variable	Saturate	P	Uncompetitive
Subsaturate	Variable	P	Mixed type
Saturate	Variable	P	Mixed type
Variable	Subsaturate	Q	Uncompetitive
Variable	Saturate	Q	Uncompetitive
Subsaturate	Variable	Q	Uncompetitive
Saturate	Variable	Q	Uncompetitive
Variable	Subsaturate	R	Competitive
Variable	Saturate	R	Competitive
Subsaturate	Variable	R	Mixed type
Saturate	Variable	R	No inhibition

Chapter 7

7.1 The product inhibition pattern is consistent with the following reaction sequence.

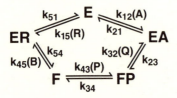

Fig. V.7.1. Reaction sequence for a uni-bi, uni-uni, ping-pong reaction.

7.2 The product inhibition pattern is consistent with either of the following reaction sequences in V. 7.2.

Chapter 8

8.1a The data are consistent with the following reaction sequence Figure V. 8.1.

8.1b In the absence of substrate B, there would be no way to convert enzyme species F to FQ if Q were not present. All of the enzyme would accumulate as species F, and F does not participate in the A → P exchange.

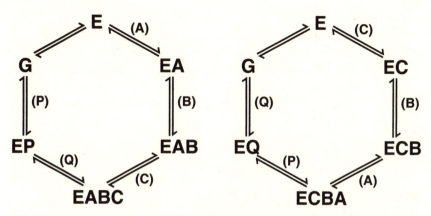

Fig. V.7.2. Two possible reaction sequences for a ter-bi enzymic reaction.

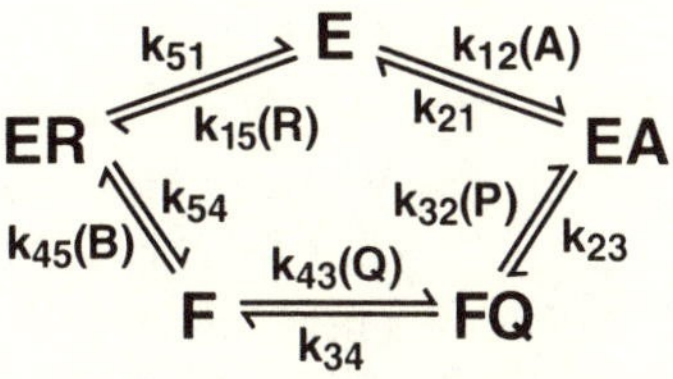

Fig. V.8.1. Reaction sequence of a uni-bi, uni-uni, ping-pong reaction.

8.1c Higher concentrations of product P would not be expected to exert a profound inhibitory effect if the dissociation of products were random.

Chapter 9

9.1 $\dfrac{V_f^h}{V_f^d} = 15\left[\dfrac{0.2 + 0.1 + 15}{15(0.1 + 0.1 + 1)}\right] = 12.8$

9.2 $\dfrac{V_f^h}{V_f^d} = 15\left[\dfrac{0.1 + 0.1 + 15}{15(0.1 + 0.1 + 1)}\right] = 12.7$

9.3 $V_f = \dfrac{(2,3,4,1)E_t}{\begin{matrix}(2,0,2,1)\\(2,0,4,1)\\(2,3,0,1)\\(2,3,4,0)\end{matrix}} = \dfrac{k_{23}k_{34}k_{41}E_t}{k_{32}k_{41} + k_{34}k_{41} + k_{23}k_{41} + k_{23}k_{34}}$

$$(0,1,2,1)$$
$$(0,1,4,1)$$
$$K_a = \frac{(0,3,4,1)}{(2,0,2,1)} = \frac{k_{41}(k_{21}k_{32} + k_{21}k_{34} + k_{23}k_{34})}{k_{12}(k_{32}k_{41} + k_{34}k_{41} + k_{23}k_{41} + k_{23}k_{34})}$$
$$(2,0,4,1)$$
$$(2,3,0,1)$$
$$(2,3,4,0)$$

$$\frac{V_f}{K_a} = \frac{k_{12}k_{23}k_{34}E_t}{k_{21}k_{32} + k_{21}k_{34} + k_{23}k_{34}}$$

$$\frac{(V_f/K_a)^h}{(V_f/K_a)^d} = \frac{k_{23}^h}{k_{23}^d}\left[\frac{1 + \dfrac{k_{32}^d}{k_{34}} + \dfrac{k_{23}^d}{k_{21}}}{1 + \dfrac{k_{32}^h}{k_{34}} + \dfrac{k_{23}^h}{k_{21}}}\right]$$

Chapter 10

10.1 Apparent K_m at pH 8.7 is $12.74\,\mu M$. Apparent V_{max} at pH 8.7 is $0.1667\,\mu$moles per minute. The apparent K_m at pH 8.2 is $33.84\,\mu M$ while the apparent V_{max} at pH 8.2 is $0.1667\,\mu$moles per minute.

10.2 Both sets of data fall on the same line, the K_m is $2.067\,\mu M$ while the V_{max} is $0.1613\,\mu$moles per minute. This suggests that the true substrate for the enzyme is the base.

$$\frac{(E^n)}{E_t} \qquad \frac{(EA^n)}{E_t} \qquad \frac{(F^n)}{E_t} \qquad \frac{(EQ^n)}{E_t}$$

10.3 $\tilde{K}_a(B)f_e^n \qquad (A)(B)f_{ea}^n \qquad \tilde{K}_b(A)f_f^n \qquad (A)(B)f_{eq}^n$

$$\frac{\tilde{K}_{ia}\tilde{K}_b}{\tilde{K}_{ip}}(P)f_e^n \quad \frac{\tilde{K}_b}{\tilde{K}_{ip}}(A)(P)f_{ea}^n \quad \frac{\tilde{K}_a\tilde{K}_{ib}}{\tilde{K}_{iq}}(Q)f_f^n \quad \frac{\tilde{K}_a}{\tilde{K}_{iq}}(B)(Q)f_{eq}^n$$

$$v = \cfrac{\cfrac{\tilde{V}_f}{f_{ea}^n + f_{eq}^n}}{\begin{array}{l}1 + \dfrac{\tilde{K}_a}{(A)}\left(\dfrac{f_e^n}{f_{ea} + f_{eq}}\right) + \dfrac{\tilde{K}_b}{(B)}\left(\dfrac{f_f^n}{f_{ea}^n + f_{eq}^n}\right) \\[3mm] + \dfrac{\tilde{K}_b}{\tilde{K}_{ip}}\dfrac{(P)}{(B)}\left(\dfrac{f_{ea}^n}{f_{ea}^n + f_{eq}^n}\right) + \dfrac{\tilde{K}_a}{\tilde{K}_{iq}}\dfrac{(Q)}{(A)}\left(\dfrac{f_{eq}^n}{f_{ea}^n + f_{eq}^n}\right) \\[3mm] + \dfrac{\tilde{K}_{ia}\tilde{K}_b}{\tilde{K}_{ip}}\dfrac{(P)}{(A)(B)}\left(\dfrac{f_e^n}{f_{ea}^n + f_{eq}^n}\right) + \dfrac{\tilde{K}_a\tilde{K}_{ib}}{\tilde{K}_{iq}}\dfrac{(Q)}{(A)(B)}\left(\dfrac{f_f^n}{(f_{ea}^n + f)eq^n}\right)\end{array}}$$

Chapter 11

11.1 The equation would be a 2:2 rational function.

11.2 $v = \dfrac{V_1(A)}{K_1+(A)} + \dfrac{V_2(A)}{K_2+(A)} + \dfrac{V_3(A)}{K_3+(A)}$

$$v = \frac{\begin{aligned}&[V_1 K_2 K_3 + V_2 K_1 K_3 + V_3 K_1 K_2](A) \\ &\quad + [V_1(K_2+K_3) + V_2(K_1+K_3) + V_3(K_1+K_2)](A)^2 \\ &\quad + [V_1+V_2+V_3](A)^3\end{aligned}}{\begin{aligned}&K_1 K_2 K_3 + [K_1 K_2 + K_1 K_3 + K_2 K_3](A) \\ &\quad + [K_1+K_2+K_3](A)^2 + (A)^3\end{aligned}}$$

Chapter 12

12.1 The condition which must be met if the substrate-saturation curve is to be sigmoidal is $\alpha_2\beta_0 > \alpha_1\beta_1$.

$(V_1+V_2)K_1 K_2 > (V_1 K_2 + V_2 K_1)(K_1+K_2)$

$0 > V_1 K_2^2 + V_2 K_1^2$

Since all of the steady state parameters are positive, the foregoing requirement cannot be met, and the equation cannot describe a sigmoidal substrate-saturation curve.

12.2 The condition which must be met if the substrate-saturation curve is to be hyperbolic is

$\beta_1 = \dfrac{\alpha_2}{\alpha_1}\beta_0 + \dfrac{\alpha_1}{\alpha_2}\beta_2$

$K_1^2 + K_2^2 = 2K_1 K_2$

This condition can be met only if $K_1 = K_2$.

12.3 From the foregoing, it can be seen that the substrate-saturation curve will be characteristic of negative cooperativity if the Michaelis constants of the two enzymes are *not* equal.

12.4 The condition which must met if substrate inhibition is to be observed is

$\alpha_1\beta_2 > \alpha_2\beta_1$

$0 > V_1 K_1 + V_2 K_2$

Therefore, the equation cannot describe substrate inhibition.

Chapter 13

13.1 $K^* = \dfrac{k^*_{-1} + k'^*_1}{k^*_1}$, $K^{app}_1 = \dfrac{k_{-1} + k'_1}{3k_1}$

$K^{app}_2 = \dfrac{k_{-2} + k'_2}{k_2}$, $K^{app}_3 = \dfrac{3(k_{-3} + k'_3)}{k_3}$

$k_1 = \alpha^3_{ff} k^*_1$, $k_2 = \alpha_{ff} \alpha^2_{fg} k^*_1$, $k_3 = \alpha^2_{fg} \alpha_{gg} k^*_1$,

$k_{-1} = \alpha_{ff} \alpha^2_{fg} k^*_{-1}$, $k_{-2} = \alpha^2_{fg} \alpha_{gg} k^*_{-1}$, $k_{-3} = \alpha^3_{gg} k^*_{-1}$,

$k'_1 = \alpha_{ff} \alpha^2_{fg} k'^*_1$, $k'_2 = \alpha^2_{fg} \alpha_{gg} k'^*_1$, $k'_3 = \alpha^3_{gg} k'^*_1$,

$K^{app}_1 = \dfrac{\alpha^2_{fg} K^*}{3\alpha^2_{ff}}$, $K^{app}_2 = \dfrac{\alpha_{gg} K^*}{\alpha_{ff}}$, $K^{app}_3 = \dfrac{3\alpha^2_{gg} K^*}{\alpha^2_{fg}}$,

$E_t = (f_3) + (f_2 gA) + (fg_2 A_2) + (g_3 A_3)$

$(f_2 gA) = \dfrac{3\alpha^2_{ff}(A)}{\alpha^2_{fg} K^*}(f_3)$

$(f_2 A_2) = \dfrac{3\alpha^2_{ff}(A)^2}{\alpha^2_{fg} \alpha_{gg} K^{*2}}(f_3)$

$(g_3 A_3) = \dfrac{\alpha^3_{ff}(A)^3}{\alpha^3_{gg} K^{*3}}(f_3)$

$(f_3) = \dfrac{K^{*3} E_t}{K^{*3} + \dfrac{3\alpha^2_{ff} K^{*2}}{\alpha^2_{fg}}(A) + \dfrac{3\alpha^3_{ff} K^{*2}}{\alpha^2_{fg} \alpha_{gg}}(A)^2 + \dfrac{\alpha^3_{ff}}{\alpha^3_{gg}}(A)^3}$

$v = k'_1(f_2 gA) + 2k'_2(fg_2 A_2) + 3k'_3(g_3 A_3)$

$v = \dfrac{3\alpha^3_{ff} k'^*_1 E_t[K^{*2}(A) + 2K^*(A)^2 + (A)^3]}{K^{*3} + \dfrac{3\alpha^2_{ff} K^{*2}}{\alpha^2_{fg}}(A) + \dfrac{3\alpha^3_{ff} K^*}{\alpha^2_{fg} \alpha_{gg}}(A)^2 + \dfrac{\alpha^3_{ff}}{\alpha^3_{gg}}(A)^3}$

13.2 $\alpha_{ff} = \alpha_{fg} = \alpha_{gg}$

13.3 $\dfrac{2}{3} > \left(\dfrac{\alpha_{ff}}{\alpha_{fg}}\right)^2$

13.4 $\dfrac{2}{3} > \left(\dfrac{\alpha_{gg}}{\alpha_{fg}}\right)^2$

13.5 $\quad K_1^* = \dfrac{k_{-1}^* + k_1'^*}{k_1^*}, \quad K_2^* = \dfrac{k_{-2}^* + k_2'^*}{k_2^*},$

$$K_1^{app} = \frac{k_{-1} + k_1'}{3k_1}, \quad K_2^{app} = \frac{k_{-2} + k_2'}{k_2}, \quad K_3^{app} = \frac{3(k_{-3} + k_3')}{k_3},$$

$$k_1 = \alpha_{ff}^3 k_1^*, \quad k_2 = \alpha_{gg}^3 k_2^*, \quad k_1 = k_2^*,$$

$$k_{-1} = \alpha_{gg}^3 k_{-1}^*, \quad k_{-2} = \alpha_{gg}^3 k_{-2}^*, \quad k_{-3} = k_{-2}^*,$$

$$k_1' = \alpha_{gg}^3 k_1', \quad k_2' = \alpha_{gg}^3 k_2'^*, \quad k_3' = k_2',$$

$$K_1^{app} = \frac{\alpha_{gg}^3 K_1^*}{3\alpha_{ff}^3}, \quad K_2^{app} = K_2^*, \quad K_3^{app} = 3K_2^*$$

$$v = \frac{3E_t[k_1'^* K_2^{*^2}(A) + 2k_2'^* K_2^*(A)^2 + k_1'^*(A)^3]}{\left(\dfrac{\alpha_{gg}}{\alpha_{ff}}\right)^3 K_1^* K_2^* + 3K_2^{*^2}(A) + 3K_2^*(A)^2 + (A)^3}$$

13.6 $\quad \dfrac{2}{3} > \left(\alpha \dfrac{ff}{\alpha_{gg}}\right)^3 \dfrac{k_1'^* K_2^*}{k_2'^* K_1^*}$

13.7 $\quad \dfrac{2}{3} > 1$, therefore substrate inhibition is not possible.

Chapter 14

14.1

$$\begin{vmatrix} 1 & \bar{\varepsilon}_1^1 & \bar{\varepsilon}_2^1 \\ 1 & \bar{\varepsilon}_1^2 & \bar{\varepsilon}_2^2 \\ 1 & \varepsilon_1^3 & \varepsilon_2^3 \end{vmatrix} \begin{vmatrix} C_1 & C_2 & C_3 \\ C_1^{M_1} & C_2^{M_1} & C_3^{M_1} \\ C_1^{M_2} & C_2^{M_2} & C_3^{M_2} \end{vmatrix} = \begin{vmatrix} 1 & 0 & 0 \\ 0 & 1 & 0 \\ 0 & 0 & 1 \end{vmatrix}$$

14.2 $\quad C_1 + \bar{\varepsilon}_1^1 C_1^{M_1} + \bar{\varepsilon}_2^1 C_1^{M_2} = 1$
$\qquad C_1 - \varepsilon_1^2 C_1^{M_1} + \bar{\varepsilon}_2^2 C_1^{M_2} = 0$
$\qquad C_1 - \varepsilon_1^3 C_1^{M_1} + \varepsilon_2^3 C_1^{M_2} = 0$

$\qquad C_2 + \bar{\varepsilon}_1^1 C_2^{M_1} + \bar{\varepsilon}_2^1 C_2^{M_2} = 0$
$\qquad C_2 - \varepsilon_1^2 C_2^{M_1} + \bar{\varepsilon}_2^2 C_2^{M_2} = 1$
$\qquad C_2 - \varepsilon_1^3 C_2^{M_1} + \varepsilon_2^3 C_2^{M_2} = 0$

$\qquad C_3 + \bar{\varepsilon}_1^1 C_3^{M_1} + \bar{\varepsilon}_2^1 C_3^{M_2} = 0$
$\qquad C_3 - \varepsilon_1^2 C_3^{M_1} + \bar{\varepsilon}_2^2 C_3^{M_2} = 0$
$\qquad C_3 - \varepsilon_1^3 C_3^{M_1} + \varepsilon_2^3 C_3^{M_2} = 1$

Flux control coefficient summation theorem.

$$C_1 + C_2 + C_3 = 1$$

Concentration control summation theorem.

$$C_1^{M_1} + C_2^{M_1} + C_3^{M_1} = 0$$
$$C_1^{M_2} + C_2^{M_2} + C_3^{M_2} = 0$$

Flux control connectivity theorem.

$$\bar{\varepsilon}_1^1 C_1 - \varepsilon_1^2 C_2 - \varepsilon_1^3 C_3 = 0$$
$$\bar{\varepsilon}_2^1 C_1 + \bar{\varepsilon}_2^2 C_2 - \varepsilon_2^3 C_3 = 0$$

Concentration control connectivity theorem.

$$\bar{\varepsilon}_1^1 C_1^{M_1} - \varepsilon_1^2 C_2^{M_1} + \varepsilon_1^3 C_3^{M_1} = 1$$
$$\bar{\varepsilon}_1^1 C_1^{M_2} - \varepsilon_1^2 C_2^{M_2} + \varepsilon_1^3 C_3^{M_2} = 0$$
$$\bar{\varepsilon}_2^1 C_1^{M_1} - \bar{\varepsilon}_2^2 C_2^{M_1} + \varepsilon_2^3 C_3^{M_1} = 0$$
$$\bar{\varepsilon}_2^1 C_1^{M_2} - \bar{\varepsilon}_2^2 C_2^{M_2} + \varepsilon_2^3 C_3^{M_2} = 1$$

14.3 $|D| = \varepsilon_1^2 \varepsilon_2^3 + \varepsilon_1^3 \bar{\varepsilon}_2^2 + \bar{\varepsilon}_1^1 \varepsilon_2^3 - \varepsilon_1^3 \bar{\varepsilon}_2^1 + \bar{\varepsilon}_1^1 \bar{\varepsilon}_2^2 + \varepsilon_1^2 \bar{\varepsilon}_2^1$

Flux control coefficients

$$C_1 = (\varepsilon_1^2 \varepsilon_2^3 + \varepsilon_1^3 \bar{\varepsilon}_2^2)/|D|$$
$$C_2 = (\bar{\varepsilon}_1^1 \varepsilon_2^3 - \varepsilon_1^3 \bar{\varepsilon}_2^1)/|D|$$
$$C_3 = (\bar{\varepsilon}_1^1 \bar{\varepsilon}_2^2 + \varepsilon_1^2 \bar{\varepsilon}_2^1)/|D|$$

Concentration control coefficients.

$$C_1^{M_1} = (\varepsilon_2^3 + \bar{\varepsilon}_2^2)/|D|$$
$$C_2^{M_1} = -(\varepsilon_2^3 + \bar{\varepsilon}_2^1)/|D|$$
$$C_3^{M_1} = -(\bar{\varepsilon}_2^2 - \bar{\varepsilon}_2^1)/|D|$$

$$C_1^{M_2} = (\varepsilon_1^2 - \varepsilon_1^3)/|D|$$
$$C_2^{M_2} = (\bar{\varepsilon}_1^1 + \varepsilon_1^3)/|D|$$
$$C_3^{M_2} = -(\varepsilon_1^2 + \bar{\varepsilon}_1^1)/|D|$$

14.4 $|D| = -[\varepsilon_1^2 \varepsilon_2^3 \varepsilon_3^4 + \bar{\varepsilon}_1^1 \varepsilon_2^3 \varepsilon_3^4 + \bar{\varepsilon}_1^1 \bar{\varepsilon}_2^2 \varepsilon_3^4 + \bar{\varepsilon}_1^1 \bar{\varepsilon}_2^2 \bar{\varepsilon}_3^3$
$\qquad + \varepsilon_1^2 \bar{\varepsilon}_2^1 \varepsilon_3^4 + \varepsilon_1^2 \bar{\varepsilon}_2^1 \bar{\varepsilon}_3^3 + \varepsilon_1^2 \varepsilon_2^3 \bar{\varepsilon}_3^1]$

Flux control coefficients.

$$C_1 = \varepsilon_1^2 \varepsilon_2^3 \varepsilon_3^4/|D|$$
$$C_2 = \bar{\varepsilon}_1^1 \varepsilon_2^3 \varepsilon_3^4/|D|$$
$$C_3 = (\bar{\varepsilon}_1^1 \bar{\varepsilon}_2^2 + \varepsilon_1^2 \bar{\varepsilon}_2^1) \varepsilon_3^4/|D|$$
$$C_4 = [\bar{\varepsilon}_1^1 \bar{\varepsilon}_2^2 \bar{\varepsilon}_3^3 + \varepsilon_1^2 (\bar{\varepsilon}_2^1 \bar{\varepsilon}_3^3 + \varepsilon_2^3 \bar{\varepsilon}_3^1)]/|D|$$

Concentration control coefficients.

$$C_1^{M_1} = [(\varepsilon_2^3 + \bar{\varepsilon}_2^2)\varepsilon_3^4 + \bar{\varepsilon}_2^2\bar{\varepsilon}_3^3]/|\mathbf{D}|$$
$$C_2^{M_1} = -[\varepsilon_2^3(\varepsilon_3^4 + \bar{\varepsilon}_3^1) + \bar{\varepsilon}_2^2(\varepsilon_3^4 + \bar{\varepsilon}_3^3)]/|\mathbf{D}|$$
$$C_3^{M_1} = -[\bar{\varepsilon}_2^2(\varepsilon_3^4 + \bar{\varepsilon}_3^1) - \bar{\varepsilon}_2^1\varepsilon_3^4]/|\mathbf{D}|$$
$$C_4^{M_1} = -[\bar{\varepsilon}_2^2\bar{\varepsilon}_3^3 - \bar{\varepsilon}_2^1\bar{\varepsilon}_3^3 - (\varepsilon_2^3 + \bar{\varepsilon}_2^2)\varepsilon_3^1]/|\mathbf{D}|$$

$$C_1^{M_2} = \varepsilon_1^2(\varepsilon_3^4 + \bar{\varepsilon}_3^3)/|\mathbf{D}|$$
$$C_2^{M_2} = \bar{\varepsilon}_1^1(\varepsilon_3^4 + \bar{\varepsilon}_3^3)/|\mathbf{D}|$$
$$C_3^{M_2} = -[(\varepsilon_1^2 + \bar{\varepsilon}_1^1)\varepsilon_3^4 + \varepsilon_1^2\bar{\varepsilon}_3^1]/|\mathbf{D}|$$
$$C_4^{M_2} = -[(\varepsilon_1^2 + \bar{\varepsilon}_1^1)\bar{\varepsilon}_3^3 - \varepsilon_1^2\bar{\varepsilon}_3^1]/|\mathbf{D}|$$

$$C_1^{M_3} = \varepsilon_1^2\bar{\varepsilon}_2^3/|\mathbf{D}|$$
$$C_2^{M_3} = \bar{\varepsilon}_1^1\varepsilon_2^3/|\mathbf{D}|$$
$$C_3^{M_3} = (\bar{\varepsilon}_1^1\bar{\varepsilon}_2^2 + \varepsilon_1^2\bar{\varepsilon}_2^1)/|\mathbf{D}|$$
$$C_4^{M_3} = -[\varepsilon_1^2\varepsilon_2^3 + \bar{\varepsilon}_1^1\varepsilon_2^3 + \bar{\varepsilon}_1^1\bar{\varepsilon}_2^2 + \varepsilon_1^2\bar{\varepsilon}_2^1]/|\mathbf{D}|$$

Chapter 15

$$C_1^1 = \hat{J}_2 C_1^2 + \hat{J}_3 C_1^3 = -\bar{\varepsilon}_1^1 C_1^{M_1} + \pi_1^1$$
$$C_{2.1}^2 = \varepsilon_1^{2.1} C_{2.1}^{M_1} + \varepsilon_2^{2.1} C_{2.1}^{M_2} + \pi_{2.1}^{2.1}$$
$$C_2^3 = \varepsilon_1^2 C_2^{M_1} - \bar{\varepsilon}_2^2 C_2^{M_2} + \pi_2^2$$
$$C_3^3 = \varepsilon_2^3 C_3^{M_2} + \pi_3^3$$

$$\begin{vmatrix} \hat{J}_2 & \hat{J}_3 & \bar{\varepsilon}_1^1 & 0 \\ 1 & 0 & -\varepsilon_1^{2.1} & -\varepsilon_2^{2.1} \\ 0 & 1 & -\varepsilon_1^2 & \bar{\varepsilon}_2^2 \\ 0 & 1 & 0 & -\varepsilon_2^3 \end{vmatrix} \begin{vmatrix} C_1^2 & C_{2.1}^2 & C_2^2 & C_3^2 \\ C_1^3 & C_{2.1}^3 & C_2^3 & C_3^3 \\ C_1^{M_1} & C_{2.1}^{M_1} & C_2^{M_1} & C_3^{M_1} \\ C_1^{M_2} & C_{2.1}^{M_2} & C_2^{M_2} & C_3^{M_2} \end{vmatrix} = \begin{vmatrix} 1 & 0 & 0 & 0 \\ 0 & 1 & 0 & 0 \\ 0 & 0 & 1 & 0 \\ 0 & 0 & 0 & 1 \end{vmatrix}$$

15.2
$$\hat{J}_2 C_1^2 + \hat{J}_3 C_1^3 + \bar{\varepsilon}_1^1 C_1^{M_1} = 1 \qquad \hat{J}_2 C_{2.1}^2 + \hat{J}_3 C_{2.1}^3 + \bar{\varepsilon}_1^1 C_{2.1}^{M_1} = 0$$
$$C_1^2 - \varepsilon_1^{2.1} C_1^{M_1} - \varepsilon_2^{2.1} C_1^{M_2} = 0 \qquad C_{2.1}^2 - \varepsilon_1^{2.1} C_{2.1}^{M_1} - \varepsilon_2^{2.1} C_{2.1}^{M_2} = 1$$
$$C_1^3 - \varepsilon_1^2 C_1^{M_1} + \bar{\varepsilon}_2^2 C_1^{M_2} = 0 \qquad C_{2.1}^3 - \varepsilon_1^2 C_{2.1}^{M_1} + \bar{\varepsilon}_2^2 C_{2.1}^{M_2} = 0$$
$$C_1^3 - \varepsilon_2^3 C_1^{M_2} = 0 \qquad C_{2.1}^3 - \varepsilon_2^3 C_{2.1}^{M_2} = 0$$

$$\hat{J}_2 C_2^2 + \hat{J}_3 C_2^3 + \bar{\varepsilon}_1^1 C_2^{M_1} = 0 \qquad \hat{J}_2 C_3^2 + \hat{J}_3 C_3^3 + \bar{\varepsilon}_1^1 C_3^{M_1} = 0$$
$$C_2^2 - \varepsilon_1^{2.1} C_2^{M_1} - \varepsilon_2^{2.1} C_2^{M_2} = 0 \qquad C_3^2 - \varepsilon_1^{2.1} C_3^{M_1} - \varepsilon_2^{2.1} C_3^{M_2} = 0$$
$$C_2^3 - \varepsilon_1^2 C_2^{M_1} + \bar{\varepsilon}_2^2 C_2^{M_2} = 1 \qquad C_3^3 - \varepsilon_1^2 C_3^{M_1} + \bar{\varepsilon}_2^2 C_3^{M_2} = 0$$
$$C_2^3 - \varepsilon_2^3 C_2^{M_2} = 0 \qquad C_3^3 - \varepsilon_2^3 C_3^{M_2} = 1$$

Flux control summation theorem.

$$\hat{J}_2 C_1^2 + C_{2.1}^2 = 1 \qquad \hat{J}_2 C_1^3 + C_{2.1}^3 = 0$$
$$\hat{J}_3 C_1^2 + C_2^2 + C_3^2 = 0 \qquad \hat{J}_3 C_1^3 + C_2^3 + C_3^3 = 1$$

Concentration control summation theorem.

$$\hat{J}_2 C_1^{M_1} + C_{2.1}^{M_1} = 0 \qquad \hat{J}_2 C_1^{M_2} + C_{2.1}^{M_2} = 0$$
$$\hat{J}_3 C_1^{M_1} + C_2^{M_1} + C_3^{M_1} = 0 \qquad \hat{J}_3 C_1^{M_2} + C_2^{M_2} + C_3^{M_2} = 0$$

Flux control connectivity theorem.

$$\bar{\varepsilon}_1^1 C_1^2 - \varepsilon_1^{2.1} C_{2.1}^2 - \varepsilon_1^2 C_2^2 = 0 \qquad -\varepsilon_2^{2.1} C_{2.1}^2 + \bar{\varepsilon}_2^2 C_2^2 - \varepsilon_2^3 C_3^2 = 0$$
$$\bar{\varepsilon}_1^1 C_1^3 - \varepsilon_1^{2.1} C_{2.1}^3 - \varepsilon_1^2 C_2^3 = 0 \qquad -\varepsilon_2^{2.1} C_{2.1}^3 + \bar{\varepsilon}_2^2 C_2^3 - \varepsilon_2^3 C_3^3 = 0$$

Concentration control connectivity theorem.

$$\bar{\varepsilon}_1^1 C_1^{M_1} - \varepsilon_2^{2.1} C_{2.1}^{M_1} - \varepsilon_1^2 C_2^{M_1} = 1 \qquad -\varepsilon_2^{2.1} C_{2.1}^{M_1} + \bar{\varepsilon}_2^2 C_2^{M_1} - \varepsilon_2^3 C_3^{M_1} = 0$$
$$\bar{\varepsilon}_1^1 C_1^{M_2} - \varepsilon_2^{2.1} C_{2.1}^{M_2} - \varepsilon_1^2 C_2^{M_2} = 0 \qquad -\varepsilon_2^{2.1} C_{2.1}^{M_2} + \bar{\varepsilon}_2^2 C_2^{M_2} - \varepsilon_2^3 C_3^{M_2} = 1$$

15.3 $\quad |D| = -[\hat{J}_2(\varepsilon_1^{2.1}(\varepsilon_2^3 + \bar{\varepsilon}_2^2) + \varepsilon_1^2 \varepsilon_2^{2.1}) + \hat{J}_3 \varepsilon_1^2 \varepsilon_2^3 + \bar{\varepsilon}_1^1(\varepsilon_2^3 + \varepsilon_2^2)]$

$$C_1^2 = [\varepsilon_1^{2.1}(\varepsilon_2^3 + \bar{\varepsilon}_2^2) + \varepsilon_1^2 \varepsilon_2^{2.1}]/|D|$$
$$C_{2.1}^2 = [\hat{J}_3 \varepsilon_1^2 \varepsilon_2^3 + \bar{\varepsilon}_1^1(\varepsilon_2^3 + \varepsilon_2^2)]/|D|$$
$$C_2^2 = -[\hat{J}_3 \varepsilon_1^{2.1} \varepsilon_2^3 - \bar{\varepsilon}_1^1 \varepsilon_2^{2.1}]/|D|$$
$$C_3^2 = -[\hat{J}_3(\varepsilon_2^3 \bar{\varepsilon}_3^2 + \varepsilon_1^2 \varepsilon_2^{2.1}) + \bar{\varepsilon}_1^1 \varepsilon_2^{2.1}]/|D|$$

$$C_1^3 = \varepsilon_1^2 \varepsilon_2^3/|D|$$
$$C_{2.1}^3 = -\hat{J}_2 \varepsilon_1^2 \varepsilon_2^3/|D|$$
$$C_2^3 = [(\hat{J}_2 \varepsilon_1^{2.1} + \bar{\varepsilon}_1^1)\varepsilon_2^3]/|D|$$
$$C_3^3 = [\hat{J}_2(\varepsilon_1^{2.1} \bar{\varepsilon}_2^2 + \varepsilon_1^2 \varepsilon_2^{2.1}) + \bar{\varepsilon}_1^1 \bar{\varepsilon}_2^2]/|D|$$

$$C_1^{M_1} = (\varepsilon_2^3 + \bar{\varepsilon}_2^2)/|D| \qquad\qquad C_1^{M_2} = \varepsilon_1^2|D|$$
$$C_{2.1}^{M_1} = -\hat{J}_2(\varepsilon_2^3 + \bar{\varepsilon}_2^2)/|D| \qquad C_{2.1}^{M_2} = -\hat{J}_2 \varepsilon_1^2/|D|$$
$$C_2^{M_1} = -(\hat{J}_2 \varepsilon_2^{2.1} + \hat{J}_3 \varepsilon_2^3)/|D| \qquad C_2^{M_2} = (\hat{J}_2 \varepsilon_1^{2.1} + \bar{\varepsilon}_1^1)/|D|$$
$$C_3^{M_1} = -(\hat{J}_3 \bar{\varepsilon}_2^2 - \hat{J}_2 \varepsilon_2^{2.1})/|D| \qquad C_3^{M_2} = -[\hat{J}_2 \varepsilon_1^{2.1} + \hat{J}_3 \varepsilon_1^2 + \bar{\varepsilon}_1^1]/|D|$$

15.4 $\quad C_1^1 = \hat{J}_2 C_1^2 + \hat{J}_4 C_1^4 + \hat{J}_5 C_1^5 = -\bar{\varepsilon}_1^1 C_1^{M_1} + \pi_1^1$
$$C_{2.1}^2 = \varepsilon_1^{2.1} C_{2.1}^{M_1} + \pi_{2.1}^{2.1}$$
$$C_2^3 = \hat{J}_4 C_2^4 + \hat{J}_5 C_2^5 = \varepsilon_1^2 C_2^{M_1} - \bar{\varepsilon}_2^2 C_2^{M_2} + \pi_2^2$$
$$C_{4.1}^4 = \varepsilon_2^{4.1} C_{4.1}^{M_2} + \pi_{4.1}^{4.1}$$
$$C_3^5 = \varepsilon_2^3 C_3^{M_2} + \pi_3^3$$

$$\begin{vmatrix} \hat{J}_2 & \hat{J}_4 & \hat{J}_5 & \bar{\varepsilon}_1^1 & 0 \\ 1 & 0 & 0 & -\varepsilon_1^{2.1} & 0 \\ 0 & \hat{J}_4 & \hat{J}_5 & -\varepsilon_1^2 & \bar{\varepsilon}_2^2 \\ 0 & 1 & 0 & 0 & -\varepsilon_2^{4.1} \\ 0 & 0 & 1 & 0 & -\varepsilon_2^3 \end{vmatrix} \begin{Vmatrix} C_1^2 & C_{2.1}^2 & C_2^2 & C_{4.1}^2 & C_3^2 \\ C_1^4 & C_{2.1}^4 & C_2^4 & C_{4.1}^4 & C_3^4 \\ C_1^5 & C_{2.1}^5 & C_2^5 & C_{4.1}^5 & C_3^5 \\ C_1^{M_1} & C_{2.1}^{M_1} & C_2^{M_1} & C_{4.1}^{M_1} & C_3^{M_1} \\ C_1^{M_2} & C_{2.1}^{M_2} & C_2^{M_2} & C_{4.1}^{M_2} & C_3^{M_2} \end{Vmatrix}$$

$$= \begin{vmatrix} 1 & 0 & 0 & 0 & 0 \\ 0 & 1 & 0 & 0 & 0 \\ 0 & 0 & 1 & 0 & 0 \\ 0 & 0 & 0 & 1 & 0 \\ 0 & 0 & 0 & 0 & 1 \end{vmatrix}$$

$$|\mathbf{D}| = -[\hat{J}_2 \varepsilon_1^{2 \cdot 1}(\bar{\varepsilon}_2^2 + \hat{J}_4 \varepsilon_2^{4 \cdot 1} + \hat{J}_5 \varepsilon_2^3) + (\varepsilon_1^2 + \bar{\varepsilon}_1^1)(\hat{J}_4 \varepsilon_2^{4 \cdot 1} + \hat{J}_5 \varepsilon_2^3) + \bar{\varepsilon}_1^1 \bar{\varepsilon}_2^2]$$

$$C_1^2 = \varepsilon_1^{2 \cdot 1}(\bar{\varepsilon}_2^2 + \hat{J}_4 \varepsilon_2^{4 \cdot 1} + \hat{J}_5 \varepsilon_2^3)/|\mathbf{D}|$$

$$C_{2 \cdot 1}^2 = [(\varepsilon_1^2 + \bar{\varepsilon}_1^1)(\hat{J}_4 \varepsilon_2^{4 \cdot 1} + \hat{J}_5 \varepsilon_2^3) + \bar{\varepsilon}_1^1 \bar{\varepsilon}_2^2]/|\mathbf{D}|$$

$$C_2^2 = -\varepsilon_1^{2 \cdot 1}(\hat{J}_4 \varepsilon_2^{4 \cdot 1} + \hat{J}_5 \varepsilon_2^3)/|\mathbf{D}|$$

$$C_{4 \cdot 1}^2 = -\hat{J}_4 \varepsilon_1^{2 \cdot 1} \bar{\varepsilon}_2^2/|\mathbf{D}|$$

$$C_3^2 = -\hat{J}_5 \varepsilon_1^{2 \cdot 1} \bar{\varepsilon}_2^2/|\mathbf{D}|$$

$$C_1^4 = \varepsilon_1^2 \varepsilon_2^{4 \cdot 1}/|\mathbf{D}|$$

$$C_{2 \cdot 1}^4 = -\hat{J}_2 \varepsilon_1^2 \varepsilon_2^{4 \cdot 1}/|\mathbf{D}|$$

$$C_2^4 = (\hat{J}_2 \varepsilon_1^{2 \cdot 1} + \bar{\varepsilon}_1^1)\varepsilon_2^{4 \cdot 1}/|\mathbf{D}|$$

$$C_{4 \cdot 1}^4 = [\hat{J}_2 \varepsilon_1^{2 \cdot 1}(\bar{\varepsilon}_2^2 + \hat{J}_5 \varepsilon_2^3) + \hat{J}_5(\varepsilon_1^2 + \bar{\varepsilon}_1^1)\varepsilon_2^3 + \bar{\varepsilon}_1^1 \bar{\varepsilon}_2^2]/|\mathbf{D}|$$

$$C_3^4 = -\hat{J}_5[\varepsilon_1^2 + \bar{\varepsilon}_1^1 + \hat{J}_2 \varepsilon_1^{2 \cdot 1}]\varepsilon_2^{4 \cdot 1}/|\mathbf{D}|$$

$$C_1^5 = \varepsilon_1^2 \varepsilon_2^2/|\mathbf{D}|$$

$$C_{2 \cdot 1}^5 = -\hat{J}_2 \varepsilon_1^2 \varepsilon_2^3/|\mathbf{D}|$$

$$C_2^5 = (\hat{J}_2 \varepsilon_1^{2 \cdot 1} + \bar{\varepsilon}_1^1)\varepsilon_2^3/|\mathbf{D}|$$

$$C_{4 \cdot 1}^5 = -\hat{J}_4(\varepsilon_1^2 + \bar{\varepsilon}_1^1 + \hat{J}_2 \varepsilon_1^{2 \cdot 1})\varepsilon_2^3/|\mathbf{D}|$$

$$C_3^5 = [\hat{J}_2 \varepsilon_1^{2 \cdot 1}(\bar{\varepsilon}_2^2 + \hat{J}_4 \varepsilon_2^{4 \cdot 1}) + \hat{J}_4(\varepsilon_1^2 + \bar{\varepsilon}_1^1)\varepsilon_2^{4 \cdot 1} + \bar{\varepsilon}_1^1 \bar{\varepsilon}_2^2]/|\mathbf{D}|$$

$$C_1^{M_1} = (\bar{\varepsilon}_2^2 + \hat{J}_4 \varepsilon_2^{4 \cdot 1} + \hat{J}_5 \varepsilon_2^3)/|\mathbf{D}| \qquad C_1^{M_2} = \varepsilon_1^2/|\mathbf{D}|$$

$$C_{2 \cdot 1}^{M_1} = -\hat{J}_2(\bar{\varepsilon}_2^2 + \hat{J}_4 \varepsilon_2^{4 \cdot 1} + \hat{J}_5 \varepsilon_2^3)/|\mathbf{D}| \qquad C_{2 \cdot 1}^{M_2} = -\hat{J}_2 \varepsilon_1^2/|\mathbf{D}|$$

$$C_2^{M_1} = -(\hat{J}_4 \varepsilon_2^{4 \cdot 1} + \hat{J}_5 \varepsilon_2^3)/|\mathbf{D}| \qquad C_2^{M_2} = (\hat{J}_2 \varepsilon_1^{2 \cdot 1} + \bar{\varepsilon}_1^1)/|\mathbf{D}|$$

$$C_{4 \cdot 1}^{M_1} = -\hat{J}_4 \bar{\varepsilon}_2^2/|\mathbf{D}| \qquad C_{4 \cdot 1}^{M_2} = -\hat{J}_4(\varepsilon_1^2 + \bar{\varepsilon}_1^1 + \hat{J}_2 \varepsilon_1^{2 \cdot 1})/|\mathbf{D}|$$

$$C_3^{M_1} = -\hat{J}_5 \bar{\varepsilon}_2^2/|\mathbf{D}| \qquad C_3^{M_2} = -\hat{J}_5(\varepsilon_1^2 + \bar{\varepsilon}_1^1 + \hat{J}_2 \varepsilon_1^{2 \cdot 1})/|\mathbf{D}|$$

Chapter 16

16.1
$$\frac{dM_1}{dt} = \alpha_1 X_0^{\varepsilon_0^1} M_1^{-\bar{\varepsilon}_1^1} M_2^{-\bar{\varepsilon}_2^1} - \beta_1 M_1^{\varepsilon_1^2} M_2^{-\bar{\varepsilon}_2^2} = 0$$

$$\frac{dM_2}{dt} = \alpha_2 M_1^{\varepsilon_1^2} M_2^{-\bar{\varepsilon}_2^2} - \beta_2 M_1^{\varepsilon_1^3} M_2^{\bar{\varepsilon}_2^3} = 0$$

$$b_1 = \varepsilon_0^1 y_0 - (\varepsilon_1^2 + \bar{\varepsilon}_1^1)y_1 + (\bar{\varepsilon}_2^2 - \bar{\varepsilon}_2^1)y_2$$

$$b_2 = (\varepsilon_1^2 - \varepsilon_1^3)y_1 + (\varepsilon_2^3 + \bar{\varepsilon}_2^2)y_2$$

$$\begin{vmatrix} -(\varepsilon_1^2 + \bar{\varepsilon}_1^1) & (\bar{\varepsilon}_2^2 - \bar{\varepsilon}_2^1) \\ (\varepsilon_1^2 - \varepsilon_1^3) & -(\varepsilon_2^3 + \bar{\varepsilon}_2^2) \end{vmatrix} \begin{vmatrix} y_1 \\ y_2 \end{vmatrix} = \begin{vmatrix} b_1 - \varepsilon_0^1 y_0 \\ b_2 \end{vmatrix}$$

$$|\mathbf{D}| = \varepsilon_1^2 \varepsilon_2^3 + \bar{\varepsilon}_1^1 \varepsilon_2^3 + \bar{\varepsilon}_1^1 \bar{\varepsilon}_2^2 + \varepsilon_1^3 \bar{\varepsilon}_2^2 - \varepsilon_1^3 \bar{\varepsilon}_2^1 + \varepsilon_1^2 \bar{\varepsilon}_2^1$$

$$y_1 = [\varepsilon_0^1 (\varepsilon_2^3 + \bar{\varepsilon}_2^2) y_0 - (\varepsilon_2^3 + \bar{\varepsilon}_2^2) b_1 - (\bar{\varepsilon}_2^2 - \bar{\varepsilon}_2^1) b_2]/|\mathbf{D}|$$

$$y_2 = [\varepsilon_0^1 (\varepsilon_1^2 - \bar{\varepsilon}_1^3) y_0 - (\varepsilon_1^2 - \bar{\varepsilon}_1^3) b_1 - (\varepsilon_1^2 + \bar{\varepsilon}_1^1) b_2]/|\mathbf{D}|$$

16.2 $y_1 = \varepsilon_0^1 \mathbf{C}_1^{M_1} y_0 + (\mathbf{C}_2^{M_1} + \mathbf{C}_3^{M_1}) b_1 + \mathbf{C}_3^{M_1} b_2$

$\qquad y_2 = \varepsilon_0^1 \mathbf{C}_1^{M_2} y_0 + (\mathbf{C}_2^{M_2} + \mathbf{C}_3^{M_2}) b_1 + \mathbf{C}_3^{M_2} b_2$

16.3 $\dfrac{d\mathbf{M}_1}{dt} = \alpha_1 \mathbf{X}_0^{\varepsilon_0^1} \mathbf{M}_1^{-\bar{\varepsilon}_1^1} \mathbf{M}_2^{-\bar{\varepsilon}_2^1} \mathbf{M}_3^{-\bar{\varepsilon}_3^1} - \beta_1 \mathbf{M}_1^{\varepsilon_1^2} \mathbf{M}_2^{-\bar{\varepsilon}_2^2} = 0$

$\qquad \dfrac{d\mathbf{M}_2}{dt} = \alpha_2 \mathbf{M}_1^{\varepsilon_1^2} \mathbf{M}_2^{-\bar{\varepsilon}_2^2} - \beta_2 \mathbf{M}_2^{\varepsilon_2^3} \mathbf{M}_3^{-\bar{\varepsilon}_3^3} = 0$

$\qquad \dfrac{d\mathbf{M}_3}{dt} = \alpha_3 \mathbf{M}_2^{\varepsilon_2^3} \mathbf{M}_3^{-\bar{\varepsilon}_3^3} - \beta_3 \mathbf{M}_3^{\varepsilon_3^4} = 0$

$$b_1 = \varepsilon_0^1 y_0 - (\varepsilon_1^2 + \bar{\varepsilon}_1^1) y_1 + (\bar{\varepsilon}_2^2 - \bar{\varepsilon}_2^1) y_2 - \bar{\varepsilon}_3^1 y_3$$

$$b_2 = \varepsilon_1^2 y_1 - (\varepsilon_2^3 + \bar{\varepsilon}_2^2) y_2 + \bar{\varepsilon}_3^3 y_3$$

$$b_3 = \varepsilon_2^3 y_2 - (\varepsilon_3^4 + \bar{\varepsilon}_3^3) y_3$$

$$\begin{vmatrix} -(\varepsilon_1^2 + \bar{\varepsilon}_1^1) & (\bar{\varepsilon}_2^2 - \varepsilon_2^1) & -\varepsilon_3^1 \\ \varepsilon_1^2 & -(\varepsilon_2^3 + \bar{\varepsilon}_2^2) & \bar{\varepsilon}_3^3 \\ 0 & \varepsilon_2^3 & -(\varepsilon_3^4 + \bar{\varepsilon}_3^3) \end{vmatrix} \begin{vmatrix} y_1 \\ y_2 \\ y_3 \end{vmatrix} = \begin{vmatrix} b_1 - \varepsilon_0^1 y_0 \\ b_2 \\ b_3 \end{vmatrix}$$

$$|\mathbf{D}| = -\left[\begin{array}{c} \varepsilon_1^2 \varepsilon_2^3 \varepsilon_3^4 + \bar{\varepsilon}_1^1 \varepsilon_2^3 \varepsilon_3^4 + \bar{\varepsilon}_1^1 \bar{\varepsilon}_2^2 \varepsilon_3^4 + \bar{\varepsilon}_1^1 \bar{\varepsilon}_2^2 \bar{\varepsilon}_3^3 \\ + \varepsilon_1^2 \varepsilon_2^3 \bar{\varepsilon}_3^1 + \varepsilon_1^2 \bar{\varepsilon}_2^1 \varepsilon_3^4 + \varepsilon_1^2 \bar{\varepsilon}_2^1 \bar{\varepsilon}_3^3 \end{array} \right]$$

$$y_1 = \left[\begin{array}{c} \varepsilon_0^1 (\varepsilon_2^3 \varepsilon_3^4 + \bar{\varepsilon}_2^2 \varepsilon_3^4 + \bar{\varepsilon}_2^2 \bar{\varepsilon}_3^3) y_0 - (\varepsilon_2^3 \varepsilon_3^4 + \bar{\varepsilon}_2^2 \varepsilon_3^4 + \bar{\varepsilon}_2^2 \bar{\varepsilon}_3^3) b_1 \\ - [\bar{\varepsilon}_2^2 (\varepsilon_3^4 + \bar{\varepsilon}_3^3) - \bar{\varepsilon}_2^1 (\varepsilon_3^4 + \bar{\varepsilon}_3^3) - \varepsilon_2^3 \bar{\varepsilon}_3^1] b_2 \\ - [(\bar{\varepsilon}_2^2 - \varepsilon_2^1) \bar{\varepsilon}_3^3 - (\varepsilon_2^3 + \bar{\varepsilon}_2^2) \bar{\varepsilon}_3^1] b_3 \end{array} \right] \Big/ |\mathbf{D}|$$

$$y_2 = \left[\begin{array}{c} \varepsilon_0^1 \varepsilon_1^2 (\varepsilon_3^4 + \bar{\varepsilon}_3^3) y_0 - \varepsilon_1^2 (\varepsilon_3^4 + \bar{\varepsilon}_3^3) b_1 \\ - (\varepsilon_1^2 + \bar{\varepsilon}_1^1)(\varepsilon_3^4 + \bar{\varepsilon}_3^3) b_2 - [(\varepsilon_1^2 + \bar{\varepsilon}_1^1) \bar{\varepsilon}_3^3 - \varepsilon_1^2 \bar{\varepsilon}_3^1] b_3 \end{array} \right] \Big/ |\mathbf{D}|$$

$$y_3 = \left[\begin{array}{c} \varepsilon_0^1 \varepsilon_1^2 \varepsilon_2^3 y_0 - \varepsilon_1^2 \varepsilon_2^3 b_1 - (\varepsilon_1^2 + \bar{\varepsilon}_1^1) \varepsilon_2^3 b_2 \\ - [\varepsilon_1^2 (\varepsilon_2^3 + \bar{\varepsilon}_2^1) + \bar{\varepsilon}_1^1 (\varepsilon_2^3 + \bar{\varepsilon}_2^2)] b_3 \end{array} \right] \Big/ |\mathbf{D}|$$

Author index

Subject index